HISTORY OF TECHNOLOGY

HISTORY OF TECHNOLOGY

Editor

Dr Graham Hollister-Short

INSTITUTE OF HISTORICAL RESEARCH

Senate House, University of London, London WC1E 7HU

History of Technology

Volume Nineteen, 1997

Edited by
Graham Hollister-Short

MANSELL
London and Washington

First published 1997 by
Mansell Publishing Limited, *A Cassell imprint*
Wellington House, 125 Strand, London WC2R 0BB, England
PO Box 605, Herndon, VA 20172, USA

British Library Cataloguing-in-Publication Data

History of Technology – 19th volume (1997)
1. Technology – History – Periodicals

ISBN 0-7201-2365-8
ISSN 0307-5451

Typeset by BookEns Limited, Royston, Herts.
Printed and bound in Great Britain by Biddles Ltd, Guildford and King's Lynn

Contents

Editorial

We open this volume with two papers on certain technologies developed in medieval Europe. In the first of these Yves Coutant and Paul Groen investigate the invention of braking systems for windmills in a region extending from the present-day northern departments of France across the Flemish and Walloon areas of Belgium to the northern limits of Holland. Not the least interesting aspect of their account is that it raises yet again the problem of technical nomenclature. New machines, as well as the elements of which they are composed, immediately require to be given names if they are to be talked or written about by contemporaries without mind-numbing recourse to periphrastic sighting shots. This is where, for historians of technology looking for baptismal certificates, as it were, the trouble begins. A new thing may be given a variety of names. Some of these may, over time, die out. Some may change their meaning, again over time, and in the worst cases new things may sometimes be called by the names of the old things they have replaced. Long after the mechanicians, and usually also the machines they created, have disappeared, names are very often all we are left with. Colonel Hime, researching the history of gunpowder nearly a century ago, posed with great clarity one particular manifestation of the problem that terminological imprecision can produce.

> Take, for example, a word *W*, which has always been the name of a thing *M*. It is then applied to some new thing, *N*, which has been devised for the same use as *M* and answers the purpose better. *W* thus represents both *M* and *N* for an indefinite time, until *M* eventually drops into disuse, and *W* comes to mean *N*, and *N* only. The confusion necessarily arising from the equivocal meaning of *W* during this indefinite period is entirely due, of course, to [the failure of people to] coin new names for new things. If a new name had been given to *M* from the first, no difficulty would have ensued. ... But as matters have fallen out, not only have we to determine whether *W means M* or *N* when it is used during the transition period, but we have to meet the arguments of those... who insist that because *W* finally meant *N* it must have meant *N* at some bygone time when history and probability alike show that it meant *M*, and *M* only.

The second of our papers on a medieval theme focuses attention on the remarkable achievements of Nuremberg in the period 1300–1500 in developing much of the technology and, perhaps as important, the financial and accounting instruments that the efficient exploitation of a growing technological capacity demanded. But it is always the moral element that one must look for, not least in this case. When Sir Thomas Elyot was sent by King Henry VIII of England as ambassador to the court of the Holy Roman Emperor, he passed through Nuremberg. In his only surviving dispatch written from that place in March 1532 he wrote: 'Touching Nuremberg, it is the moste proper towne and best ordred publike weale that ever I beheld' – a hint of the spirit of *L'Esprit des lois* some two hundred years before it found its classic expression. The rule of law, essentially the maintenance of equity, as Wolfgang von Stromer makes abundantly clear, was the necessary foundation for Nuremberg's success. As for the idea that medieval technological development was of such range and intensity as to provoke an industrial revolution of some kind, this, I suppose, may be traced back to Carus-Wilson's essay 'An industrial revolution of the 13th century' of 1941. Since then the idea has not been well served by the publication of over-enthusiastic accounts, and has indeed been pretty well discredited under the sustained fire of its critics. Nevertheless, as further study reveals fresh evidence (such as that in this volume) of what the medieval technological achievement really amounted to, some more accurate accounting will have to be done. The magnitude of change induced in medieval society by technological innovation has in one respect been well demonstrated long since by the researches of business historians such as Raymond de Roover. The indirect effect of that change is clearly visible in the way in which the teaching of Aristotle and the doctrines of the Church came to be subjected to an enormous degree of revision so as to bring ethical teaching on the subject of trade and profits into some sort of harmonization with the actual practice of the day. This had already been substantially accomplished before the close of the thirteenth century.

If I have left myself too little space to comment on other papers in this volume, I can say at least that all of them offer fresh and stimulating insights into the nature of technological change and the constraints to which it is subject. Wilfred Lockett, in the case of Jacob Leupold, examines the important role of the *practicus/mechanicus*, as Leupold would have called himself, as a middleman making the ideas of the *theoreticus* available to the *empiricus*. Jennifer Tann, in the case of British steam engine exports to the West Indies, brings into prominence the importance of closed markets to British industry as early as the 1820s. Michael Fores puts forward the case that much of what is taken to be history of technology should really be seen as more like a kind of metahistory, a mirror of the way many writers think about the subject, rather than any very real reflection of how the subject itself might be pursued. A. P. Woolrich shows how the British patent system worked against the very people it was supposed to serve. It was a jungle through which, burning bright, John Farey made his way for the best part of forty years. His drawings illustrating this paper were done by him in 1805 when he was not yet 15 years of age. Finally, Martijn Bakker reveals a too little-researched aspect of technological diffusion (apart perhaps, that is, from the experience of the Huguenots), namely the enforced flight or prudential migration of experts seeking refuge from their political enemies.

I would like to say, in conclusion, that for making the Institute of Historical Research available as the new home of the journal and permitting it to be published under the auspices of the IHR, both the publishers and I would like to thank Professor Patrick O'Brien, Director of the Institute.

Graham Hollister-Short
London

The Contributors

Dr Martijn Bakker,
Van Slichtenhorstraat 84,
6524 Nijmegen,
The Netherlands

Dr Yves Coutant,
't Rode Paard 39,
8510 Bellegem,
Belgium

Mr Michael Fores,
80 Lexham Gardens,
London W8 5JB,
England

Mr Paul Groen,
Raadhuisplein 45,
5854 AW Bergen,
The Netherlands

Professor A. Rupert Hall,
14 Ball Lane,
Tackley,
Oxfordshire OX5 3AJ,
England

Dr Wilfred G. Lockett,
55 Wellesley Street E., Apt. 601,
Toronto,
M4Y 2T6 Canada

Dr Dennis Simms,
17 Dunstan Road,
London NW11 8AG,
England

Mr Robert D. Smith,
Head of Conservation,
The Royal Armouries,
Armouries Drive,
Leeds LS10 1LT,
England

Professor Dr Wolfgang von Stromer,
Wirtschafts- und Technikgeschichte,
Burg Grünsberg,
90518 Altdorf,
Germany

Professor Jennifer Tann,
Business School,
The University of Birmingham,
Edgbaston,
Birmingham B15 2TT,
England

Mr A. P. Woolrich,
Canal Side,
Huntworth,
Bridgwater,
Somerset TA7 0AJ,
England

Notes for Contributors

Contributions are welcome and should be sent to the editor. They are considered on the understanding that they are previously unpublished in England and are not on offer to another journal. Papers in French and German will be considered for publication, but an English summary will be required. The editor will also consider publishing English translations of papers already published in languages other than English. Three copies should be submitted, typed in double spacing (including quotations and notes) with a margin, on A4 or American Quarto paper. Include an abstract of 150–200 words and two or three sentences for 'Notes on Contributors'.

It would be appreciated if normal printers' instructions could be used. For example, words to be set in italics should be underlined and *not* put in italics. Authors who have passages originally in Cyrillic or oriental scripts should indicate the system of transliteration they have used. Quotations when long should be inset without quotation marks; when short, in single quotation marks. Spelling should follow the *Oxford English Dictionary*, and arrangement H. Hart, *Rules for Compositors* (Oxford, many editions). Be clear and consistent.

All papers should be rigorously documented, with references to primary and secondary sources typed separately from the text in double spacing and *numbered consecutively*. Cite as follows for books:

> 1. David Gooding, *Experiment and the Making of Meaning: Human Agency in Scientific Observation and Experiment* (Dordrecht, 1990), 54–5.

Scientific references may be written:

> 3. Gooding, *op. cit.* (1), 43.

Only name the publisher for good reason. For theses, cite University Microfilm order number or at least Dissertations Abstract number. Standard works like DNB, DBB must be used thus cited.

And as follows for articles:

> 13. Andrew Nahum, 'The Rotary Aero Engine', *Hist. Tech.*, 1986, 11: 125–66, p.139.

Line drawings should be drawn boldly in black ink on stout white paper, feint-ruled paper or tracing paper. Photographs should be glossy prints of good contrast and well matched for tonal range. The place of an illustration should be indicated in the margin of the text where it should also be keyed in. Each illustration must be numbered and have a caption. Xerox copies may be sent when the article is first submitted for consideration.

The Early History of the Windmill Brake

YVES COUTANT AND PAUL GROEN

INTRODUCTION

In the Middle Ages three different sources of power were available: human and animal muscle, water and wind. The first two were easy to control. At a watermill, for example, the amount of water could be regulated and turned off without any problem. Wind, however, cannot be stopped, and windmills therefore had to be provided with a braking system.

Probably the first, and still small, windmills did not have a specific braking system. R. Wailes supposed that the first small English mills were simply turned out of the direction of the wind.[1] Such a procedure is still usual in some countries. The most frequent manner of stopping the Greek tower mill is comparable with that of the Portuguese tower mill. The mill is quartered by turning it manually out of the wind. After this, the miller, staying at ground level, slows and stops the mill with a rope wound round the windshaft between the gearwheel and the neck (front journal of the windshaft).[2] In Sweden and in Brittany also, the smallest mills are stopped in roughly the same way.[3] In Finistère (Brittany), a tower mill, the moulin de Kerteodin near Cléden-Cap-Sizun, was built as late as 1756 without a brake.[4]

Small mills can also be stopped in another way: by choking the stones with grain and lowering the runner stone onto the bed stone. Most small non-orientable Greek windmills work in this way.[5]

The power output of a mill is proportional to the surface of the turning circle of the sails and thus increases with the quadratic length of the sails, and so the systems mentioned so far are not sufficient for bigger mills. Most Flemish windmills of the fourteenth and fifteenth centuries were tall. In 1451, the buck (body) of the postmill of Kaprijke (Oost-Vlaanderen, Belgium) had about the same proportions as the present Flemish postmills.[6] This means that braking systems used in that kind of Flemish mill in medieval times must have had almost the same capacity as the braking systems of mills nowadays.

The braking system of a windmill worked and still works on the gearwheel of the windshaft. In brief outline the system consists of a circular wooden band which is fastened to one end at the buck; at the other end hangs a heavy horizontal lever (Figure 1). The weight of the lever, which rotates round a pivot, provides a tensile force in the brake band. To let

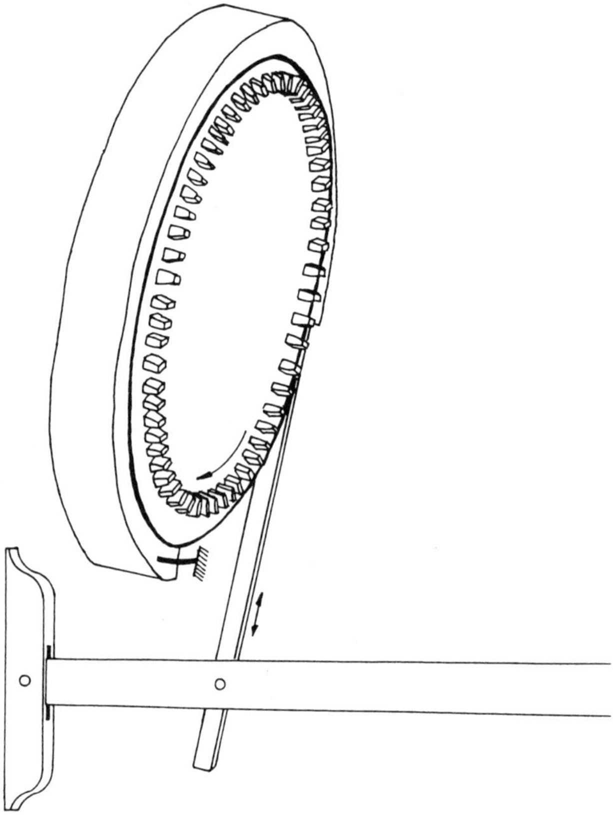

Figure 1 Outline of the braking system in a windmill.

loose the brake force and start grinding, the miller lifts the lever and hangs it in the *catch*. The principle is the same as that of a normal brake, but the action of the lever is negative: by lifting it up, the brake force is cancelled. Although the principle of such a braking system is simple, its realization is difficult. In 1410 for example, a millwright worked for six and a half days to install a new brake in the mill of Erquinghem-sur-Lys (Nord, France) .[7] In spite of modern techniques this operation takes even more time today.

THE BRAKE BAND

In the Middle Ages, two kinds of brake bands existed: a circular wooden plank and a band composed of several blocks fastened to each other.

The Circular Wooden Plank

To saw out the brake band, millwrights looked for a bent tree. Molinologist Ronse, who refers to contemporary practice of the 1930s, says millwrights left the plank under water for months to make it more flexible. After this they fixed one end to the gearwheel, fastened a rope between the ends of the sailstocks and turned the sails manually in the reverse direction. When the plank was fully bent round the wheel, the other end was fixed. Thereafter the plank was dried for a couple of days.[8] We did not find this method mentioned in the medieval records, but that does not mean it was not applied. According to the rare texts evoking the bending of the wooden brake band, millwrights used not water but fire:

> A lui que semblablement il a payé pour 1 quartier de fagotz dont les dictes retenues furent brulees et ployees pour mettre en œuvre.
> (To whom he has paid in the same way for 1 quarter of faggots for use in burning and bending of the above-mentioned brakes.)
> Algemeen Rijksarchief (Brussells), Rekenkamer 7376, 21 r° Sint-Baafs-Vijve (West-Vlaanderen, Belgium) (1450/1)

> Arent Charlet, temmerman, die heeft de pranghen gheviert, ghecromt ende in de derde (*sic*, = in derde) gheleyt.
> (Arent Charlet, carpenter, who has burned and bent the brakes and put them in the soil.)
> Rijksarchief Ghent, Sint-Pietersabdij, I 698, 27 v° Ghent (Oost-Vlaanderen, Belgium) (1475/6)

As the last sentence shows, they buried the planks in the soil to maintain their curve. In the next item we read the same:

> Item betaelt eenen man die den pit maecte omme de pranghe te doen boeghene, 14 s[cellinghen].
> (Item paid to a man who dug the pit to make the brake bend, 14 shillings.)
> Rijksarchief Ghent, Sint-Pietersabdij, I 683, 21 v° (Oost-Vlaanderen, Belgium) (1454/5)

Usually the brake band was made of elm, but willow, poplar and ash were also pressed into service. Elm and ash have the property of being tough, hard and strong. Elm is the hardest, ash the toughest. They are more difficult to work up than willow and poplar – two softer and more flexible kinds of wood – but last longer as brakes.

> A Jehan de Sceppere, pour quartre arbres omeaulx et ung fresne . . . et du fresne ont esté faiz et soiez trois retenues de molin, l'une mise et emploiee en ce dit molin, une autre au molin de Vive . . . et les autres (*sic*) sont encore en estat et de garnison . . .
> (To Jehan de Sceppere, for four elms and one ash . . . and from the ash three mill brakes have been made and sawn, one placed and used in the above-mentioned mill, another in the mill of Vive . . . and the others *(sic)* are not used and stored . . .)
> Algemeen Rijksarchief (Brussells), Rekenkamer 7383, 24 v° Ingelmunster and Sint-Eloois-Vijve (West-Vlaanderen, Belgium) (1457/8)

This excerpt shows that the brake band was an important part of the mill, and wore out rather quickly. For repairing the mill as quickly as possible they had a supply of spare brake bands. It is not impossible, however, that the *retenues* are blocks of a block brake (see below).

The Brake Band of Blocks

According to Stroop, the brake composed of blocks (Figure 2) was an improvement on the plank brake.[9] We doubt this assertion, since we found the word *blocvanghe* – the name the Flemish and the Dutch millers still used for this kind of brake band – in the repair accounts of Menin as early as 1402:

> A Sohier Vido, charpentier, pour avoir ouvré et fait un membre appellé 'blocvanghe' ou molin a vent...
> (To Sohier Vido, carpenter, for having worked and made a piece called 'blocvanghe' in the windmill...)
> Algemeen Rijksarchief (Brussells), Rolrekening 2075, 38th week

We can hardly believe that Menin was an isolated case. Two indications rather prove the opposite. The first is the kind of wood employed. Sipman supposed that willow was less suitable for manufacturing brake bands,[10] hence the use of willow would indicate that we are dealing with a block brake. The second point is the use of the plural form for the term that defines the braking system, a usage followed by some clerks. This might point to the making of spare parts, as we have encountered in the accounts of Ingelmunster and Sint-Eloois-Vijve. But often the plural refers to a single brake. Is this not the block brake, consisting of different segments? Even quite recently the miller of Moutiers-en-Beauce (Eure-et-Loir, France) used the plural *les freins* when talking about the block brake.[11] The following three examples date from the Middle Ages:

> Eeerst van eenen popelieren bole ghecocht tSwinaerde, om pranghen daer af te makene...
> (First for a poplar trunk bought at Zwijnaarde, for making brakes...)
> Rijksarchief Ghent, Sint-Pietersabdij, I 671, 22 r° Zwijnaarde (Oost-Vlaanderen, Belgium) (1427/8)

> A lui, pour une plate et lyen de fer mis et employé a l'une des retenues d'icellui molin et pour les claux y employez, 8 s[olz].
> (To him, for an iron band placed and used at one of the brakes of that mill and for the nails used, 8 shillings.)
> Algemeen Rijksarchief (Brussells), Rekenkamer 7378, 16 v° Ingelmunster (West-Vlaanderen, Belgium) (1452/3)

> Item pour avoir encore soijé 500 d'ouvrage en cartelage, dont on a fait 4 frains et pluseurs costerez pour servir au molin...
> (Item for having sawn 500 saw-cuts for making beams, for the use of 4 brakes and different uplongs (long sail laths parallel with the sailstock) at the mill...)
> Archives Départementales du Nord (Lille, France) 30 H 252, 46 r°–v° Douai (Nord, France) (1460/1)

Figure 2 'De Oude Molen', Wijchen, Gelderland (Netherlands). The blocks of the block brake can be seen round the gearwheel.

We find the plural as early as 1382 in connection with the mill of Kattendrecht (Zuid-Holland, Netherlands).[12]

> Item die molen in Kathendrecht doe die vanghe niet en hilden, dede Bouden Aerntssoen maken twe repe daer men die molen mede bant . . . (Item at the mill in Kattendrecht: as the brakes did not hold, Bouden Aerntssoen made two ropes to tie the mill . . .)

It is quite possible that the plural form used for brake in some of the above-mentioned examples points to the various parts of which the braking system is composed. The most logical explanation seems to be the use of separate brake blocks. The nails mentioned in the item of Ingelmunster also argue for blocks.

If the interpretation of these three clues (the kinds of wood used, the plural form, the nails) is correct, we must accept that the block brake was older and more widely diffused than has been thought until now and that two braking systems coexisted in the Middle Ages. The block brake needs good quality iron; it requires less time to repair than a plank brake. It has the additional advantage of lasting longer. The only drawback is the higher price. Economic criteria and personal preference would have determined the choice. Nowadays both systems still exist side by side. Sometimes the plank of the plank brake is replaced by one made of steel.

THE FASTENING

As far as we already know, the brake band can be fastened to the buck in one of three ways, according to whether the brake is a sprag brake, lip brake or Flanders brake.

The Sprag Brake

The brake band skirts between 50 and 60 per cent of the gearwheel's circumference. At the end the band has a notch at the outside. The sheer (horizontal beam supporting the cap) nearby has a similar notch. A sprag is placed in each of the two notches. When the lever is depressed, the brake band is pulled round the gearwheel, and the notch is tilted towards it. By this means the sprag presses the end of the brake band on to the gearwheel (Figure 3).

Around 1900 almost every mill on the Zealand islands still had a sprag brake, and numerous traces of this system are visible even now. In dozens of mills in Zealand, Noord-Holland and Utrecht (Netherlands) the system is maintained.[13] However, this kind of brake is not explicitly referred to in the medieval accounts.

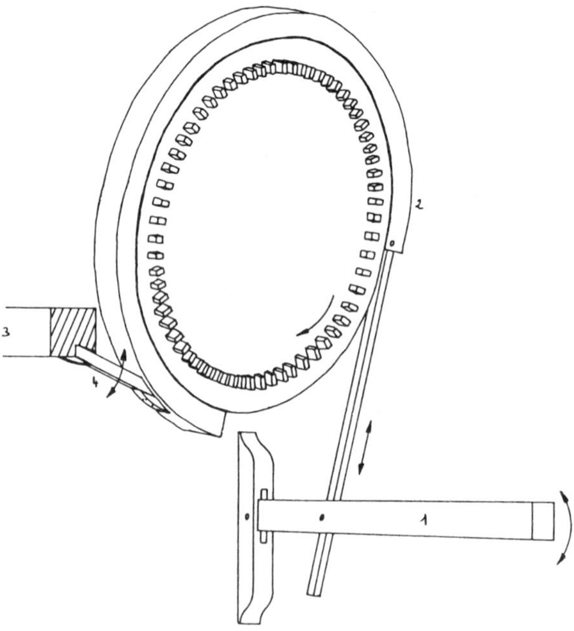

Figure 3 The sprag brake. The arrows show the movement of the different parts. 1, lever; 2, band; 3, sheer; 4, sprag.

The Lip Brake

As in the sprag brake, the brake band skirts between 50 and 60 per cent of the gearwheel's circumference. The end of the brake band is fixed to a crossbeam named 'lip'. This beam rotates at one end on a pivot. By lowering the lever, the brake band pulls up the lip so that its other end presses on the gearwheel (see Figure 4).

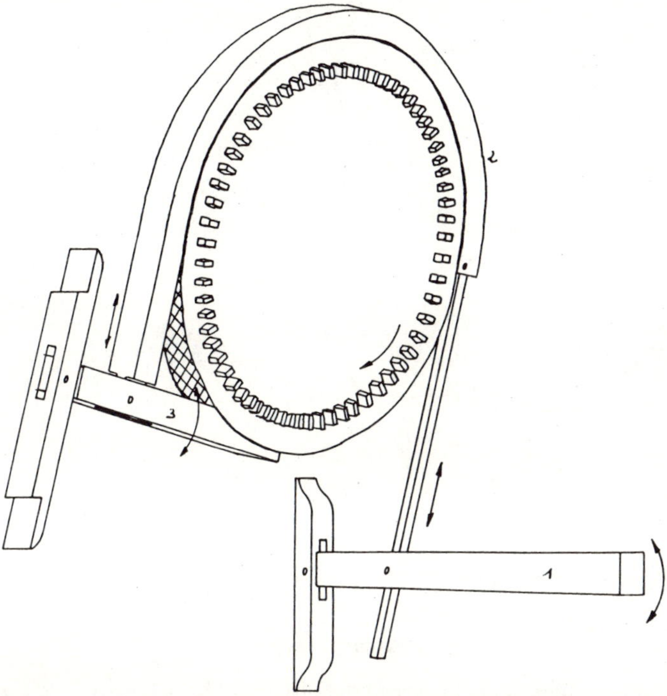

Figure 4 The lip brake. 1, lever; 2, band; 3, lip.

Nowadays the lip brake has practically disappeared throughout the whole western part of Western Europe – an exception, however, is the postmill of Dornum (Germany).[14] But lots of mills still have traces of lip brakes. Several millers have referred to a recent change in brake construction.[15] In eastern Germany and in Scandinavia, however, there are dozens of lip brakes.

The Flanders Brake

From the lever the brake band skirts almost the whole circumference of the gearwheel. The end of the brake band is fastened to the sheer with iron parts. When the lever is depressed, the brake band grips the wheel (Figure 5).

Nowadays the Flanders brake is the most widely used system in West Europe.

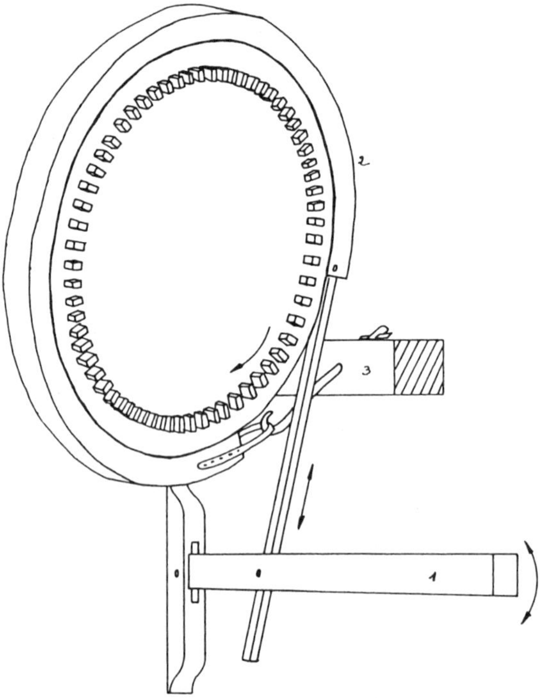

Figure 5 The Flanders brake. 1, lever; 2, band; 3, sheer.

HISTORY

Before we begin to investigate historical details, it will be useful to look at the different names of the braking systems in Flanders in the fourteenth and fifteenth centuries. These names can be arranged in three different groups:

1. The idea of checking. *Frain* is the normal name in French texts, being found in the accounts of the French part of Flanders and in some texts written by French-speaking clerks. The *frain* (brake) is a construction that bridles a movement or checks a moving thing (from the Latin *frenum*, 'bridle').
2. Terms which lay the accent on pressure. Etymology relates the Flemish word *pranghe* (or *prande*) to the Middle High German *pfrengen* and the Middle English *to prangle*, which all mean 'to press'. We find the same phenomenon in the Scandinavian and German names of the braking system in windmills. In Denmark, for instance, the braking system is called *pressen* or *persen*.[16] In German mill literature we read *Presse* and *Press-baum* (the lever).[17] Some old Flemish accounts called the brake of the windmill *parsse* or *perse*:[18]

> ... een nieuwe perseboom, een nieuwe lippe te makene ende de perse te verhanghene ...; ... van eenre niewer perse, enen niewen perseboom, 3 l [ponden] 12 s[cellinghen].
> (... to make a new *perse*-lever and a new 'lip', and to adjust the *perse* ...; ... for a new *perse*, a new *perse*-lever, 3 pounds 12 shillings.)
> Algemeen Rijksarchief (Brussels), Rolrekening 159, Praet (West-Vlaanderen, Belgium) (1376/7)

3. Terms which lay the accent on catching: *vanghe*, verbal substantive of the Dutch *vanghen* ('to catch'), and the literal translations *prise* and *retenue*, which we have found only with one receiver. Originally *vanghe* appeared only in the western part of Flanders, but during the fifteenth century the term spread to the north-east and east.

The doublet *vang–prang* (modern spelling of the medieval words *vanghe* and *pranghe*) is an essential argument in Stroop's thesis, which posits that the windmill spread from the Rhine region as well as from Normandy.[19] According to Stroop, the term *vang* was originally the only term in the present-day province of West-Vlaanderen, while *prang* was the oldest term in the present-day province of Antwerpen. During expansion to the west, the term would not have gone further than Evergem, a municipality to the north of Ghent. The word *vang* (expressing the sense of *frein*, *frain*) appeared when the windmill, having spread from Normandy, reached Flanders. *Prang*, on the other hand, would come from the Rhine region. In the course of time *vang* would gain ground on *prang*. Nowadays the word has already reached Schleswig-Holstein.

This interesting vision is basically the same as the linguistic pattern we have found in the accounts, but we will have to specify it further. First we have the Norman origin of mills. Despite the attractive and well-known theory of A.-M. Bautier,[20] there is no certainty that the first postmill was Norman. The fact that the oldest record of a postmill could be that of Viesville (Normandy) does not mean that this type of mill originated in that place. The recording of a mill's existence does not necessarily say anything about its age.[21] Let us take a more rational and less chauvinistic look at things. Since we possess only a small part of the archives from the twelfth century, we cannot know for certain whether the postmill originated on the coast of England, Normandy or Flanders. Jan Stroop suggests that the *vang* was first used when the windmill of Normandy reached Flanders. But what relation exists between the words *vang* and *frein*? None! *Vang* is a specific molinogical term and *frein* is used to mean brake in all kinds of machines. Arguing that *vang* has any relation with the Norman windmill is pure speculation. Jan Stroop based his theory too heavily on the theory of A.-M. Bautier, who systematically neglected the recordings of twelfth-century Flemish windmills. Besides, *prang* reached further west than Stroop assumed. For instance, we found the word *prang* in the accounts of Petegem-aan-de-Leie (Oost-Vlaanderen, Belgium) (1371), Harelbeke (West-Vlaanderen, Belgium) (1402/3), Sluis (Zeeland, Netherlands) (1444/5), Kaprijke (Oost-Vlaanderen, Belgium) (1452/4),

Elsegem (Oost-Vlaanderen, Belgium) (1453/4) and Petegem-aan-de-Schelde (Oost-Vlaanderen, Belgium).

Stroop did not find a recorded use of the word *vang* in West-Vlaanderen earlier than the one for Moen of 1455. We found the word in accounts of Reningelst (1374/5), Harelbeke (1378/9), Elverdinge (1398/9) and Menin (1402/3), all localities situated in the province of West-Vlaanderen. There is also the word *parsse* or *perse*, which we found in the surroundings of Bruges. In the oldest accounts of the windmill belonging to the 'Potterie' hospital of Bruges, we found only the word *parsse*. We have to wait until halfway through the fifteenth century to encounter the word *vanghe*. In the accounts of another windmill of the 'Potterie' hospital, the one of Vlissegem, the words *vanghe* and *parsse* were used interchangeably (1408/9), but here also *persse* was used more frequently. Since the words *prang* and *vang* express different ideas, they might be used for different braking systems. The *vang* is probably the braking system in which the band embraces the whole gearwheel, now called the 'Flanders *vang*' (brake). The *prang* or *perse*, which presses on the gearwheel, is probably the braking system in which the band partially embraces the gearwheel: the lip brake.

According to Stroop, the *vang* type originated in Zealand; in 1318 we encounter the word *ghevanghe* twice in accounts from Renesse, Haamstede and Arnemuiden.[22] In these two cases the context is so ambiguous that we do not know for sure if they were referring to the braking system. If *ghevanghe* was really the brake, why do we not have any other recording of the word with this meaning?[23] Stroop assumed that the prefix *ghe-* had vanished, but such a linguistic phenomenon takes place only if a word is used intensively. Nothing proves that this was the case. Another fact contradicts the theory of Stroop: people in West-Vlaanderen still call a prison *gevang* and not *vang*. Here the prefix did not vanish. The windmill possesses parts which in French accounts are called *prison*, for instance the beams whose function is to prevent the windshaft from rolling off the neck bearing and tail bearing. Even now similar constructions are found in some mills in the region of Rotterdam (see Figure 6). Does the word *ghevanghe* in the Zealand accounts have this meaning? As previously mentioned, until about 1900 basically all mills on the islands of Zealand had a sprag brake. If, as Stroop assumed, the Flanders *vang* (brake) was a discovery from Zealand, then we do not understand why the sprag brake was used until the twentieth century. Besides, accounts from Bruges – Bruges is not so far away from Zealand – do not mention the word *vanghe* before the fifteenth century. Is this not proof that the north of West-Vlaanderen and Zealand originally only knew the *perse*? Would even the Dutch millers later call the braking system in which the brake band embraces the whole gearwheel *Vlaamse vang* (Flanders *vang*) if it were an invention of Zealand?

Flemish accounts from the Middle Ages reveal another word, which has not been studied before, and which may possibly contribute to the solution of the problem: *lippe*. Let us look at the most important places where this term has been found. In the third column of Table 1 we give the name of

Figure 6 The mill 'De Doornenboom', Hilvarenbeek, Noord-Brabant (Netherlands). Note the beams which prevent the main shaft rolling off the tail bearing. We presume it was what was called the *ghevanghe* in the oldest accounts from Zealand.

the braking system, as it appears in the accounts. Most of the information is quite old. Only in Bruges and in the polders to the west of the Scheldt was the word *lippe* still used in the fifteenth century. The word *lippe* is always in relation with *perse* and *prange*, except in Elverdinge and in Bruges. *Reprinse* was probably an attempt to translate the word *vanghe* into French. In Ingelmunster, for example, the word *vanghe* was translated into *prise* and *retenue*.

All this confirms our conviction that the Flemish mills had a braking system in which the brake band embraced only a part of the gearwheel. The oldest braking system in Flanders was a *perse* or a *prange*, a lip brake. The heavy horizontal lever by which the brake band is lowered was not located on the same side as the fastening of the brake band. Nevertheless in Elverdinge and in Bruges the word is related with *vanghe*. Since language and technology do not evolve at the same pace,[24] the word *vanghe* could be used for the old lip brake. This would explain the association between *lippe* and *vanghe* in Elverdinge and in Bruges. Another account shows us that intermediate phase:

> omme twee vangelette an die parsse ...
> (for two iron parts of the brake for the brake)
> Archief van de Potterie (Bruges, Belgium), Reg. 90, 43 v°, Vlissegem (West-Vlaanderen, Belgium) (1408/9)

The reverse, the survival of the old word notwithstanding the introduction of a technical innovation, can also be observed. For example, around 1930

Table 1 Appearance of various terms in medieval records

Eeklo, Waarschot Kaprijke	1371/2	*perse*	
Saaftinge	1397/8	*prande*	
Elverdinge	1398/9	*vanghe*	une nouvelle reprinse appellee 'vanghelippe'
Saaftinge	1410/1		liphout
Verrebroek/Hulst	1425/1	*pranghe*	une autre piece de bois dont l'en fait mouldre et arrester le dit moulin, appellee en flamenc 'de lippe'...; item pour une autre piece de bois servant a la dicte lippe, 4 s par....; item a Woutre Curaert, pour l'achat et delivrance d'une piece de bois mis et employe au dit molin, ou l'en fait pendre la dessus dicte prange, 8 s par. (NB. The *prange* had not yet been mentioned in that account!)
Verrebroek	1438/9	*pranghe*	
Verrebroek	1439/40		
Bruges	1453/4	*vanghe*	
Temse	1464/5	*pranghe*	eene kenneve daer de leppe van der pranghen in hanct

old millers in the Alblasserwaard, a region 30 miles (50 km) to the east of Rotterdam (Netherlands), still called their braking system *praam*, a local equivalent of *prang*, even though the Flanders brake had probably been installed two centuries previously.[25]

Even though the accounts of Verrebroek/Hulst of 1425/6 sum up the parts of the *pranghe*, there is no mention of a metal fastening, so all the parts must have been of wood, which again proves that this mill did not have the Flanders brake.

If it was not in Zealand that the term Flanders *vang* was first used, where was it? The oldest recordings of the word *vanghe* all come from the present-day Belgian province of West-Vlaanderen. Logically it is there that we shall have to look for the first Flanders *vang*. An account from Armentiers (Nord, France) of 1412/3 gives us the following description of the braking system:

> A lui, pour le bois d'un frain fait au dit molin en lieu d'un autre qui estoit brisiés, 100 s[olz]. A lui, pour avoir fait le dit frain, c'est assavoir un cercle tout autour du rouet de 32 piés de long et d'un pié de lé, de deux paux d'espesseur, 1 flayel de 20 piés de long et une langue de 16 piés de long, 1 postiel deriere qui tient le dit flayel de 8 piés de long et de 6 paulx en quarure ... 6 l[ivres].
> (To him, for the wood of a brake that is made in the mill to replace

> another that was broken, 100 shillings. To him, for having made the above-mentioned brake, i.e. a complete circle round the gearwheel, length 32 ft, width 1 ft, thickness 2 inches, one lever which is 20 ft long and a tongue with a length of 16 ft, a stick at the end that keeps up the already-mentioned lever, length 6 ft and thickness 6 inches square ... 6 pounds.)
> Archives Départementales du Nord (Lille, France), BI4564, 81 v°

In the fifteenth century a description of technical parts was very rare, because people of the time did not do anything without reason. Mill accounts of the Middle Ages were financial documents, not technical ones. A detailed account is usually proof of a more expensive construction. The Receiver of Armentiers, one of the few who described the braking system in detail, probably had to justify a higher investment: the Flanders *vang* was more expensive than the lip brake, because the brake band was longer and the fastening necessitated heavier iron parts. Does the Receiver not emphasize the fact that the entire gearwheel was embraced by the brake band (*tout autour du rouet*)? The brake band measured approximately 9.50 × 0.29 × 0.05 m. It is noteworthy that the brake band as described in the *Encyclopédie* of Diderot and d'Alembert was also a wooden plank made of elm and 32 ft (9.50 m) long. The lever (*flayel*) is approximately 5.80 m long, which should be the approximate depth of the buck. The *langue* would be the extension piece by which the brake band was connected to the lever (which was probably located in the meal floor underneath the stones).[26] The *langue* was around 4.65 m long.[27]

We are convinced of the fact that the Flanders brake originated in the south of West-Vlaanderen or the north of France in the fourteenth century. The Armentiers text shows that it was not in common use at the beginning of the fifteenth century. The new braking system began to spread, albeit very slowly at first, because despite its greater cost it was more effective and possibly had a longer working life. This clarifies the substitution of *pranghe* by *vanghe*.

Not until 1475 did the Flanders *vang* reach Ghent:

> Paeschier Karin, smedt, die heeft een ooghe met twee vederen ende eenen haec met grooten cricken an de voorseide pranghe gheslegghen om te slutene *up een nieu maniere* [our emphasis] ...
> (Paeschier Karin, blacksmith, who has fastened an eye, with two feathers and a hook with big cranks to the above-mentioned brake to fix it *in a new way* ... [see Figure 5])
> Rijksarchief Ghent, St-Pietersabdij, I 698, 27 v°, Ghent (Oost-Vlaanderen, Belgium) (1476/7)

> Paeschier Karin, smedt, die heeft ghemaect eene ooghe met twee langhe vederen, elke vedere van 6 voeten lanc, omme de selve pranghe, ende oic eenen haec met eender crucke ...
> (Paeschier Karin, blacksmith, who has made an eye with two long feathers, each feather 6 ft long, for the same brake, and also a hook with a crank ...)

> Rijksarchief Ghent, St-Pietersabdij, I 698, 25 v°, Zwijnaarde (Oost-Vlaanderen, Belgium) (1477/8)

As soon as the millwright had recognized the value of the new brake, he replaced the braking system of the two mills he had to maintain.

In 1522 we read in the accounts of Axel (Zealand, Netherlands) that the local mill was equipped with a *houppranghe*, fastened by various iron parts. The word *houppranghe* is an outstanding example of the fact that language evolves more slowly than technology. The first part, *houp* ('hoop'), means that the brake band embraces the whole wheel. The second part, *pranghe*, is the name of the original braking system.

> Itemin in den eerst ghecocht te Vinckele teghen Massijons een houppranghe, de welke coste 3 s[chillingen] 4 g[roten]. Item betalt Joos de Soemer ende Jan Balsan van der selver houppranghe in te doene ghelicht sij behort, 2 s[chillingen] groot ... Item eerst gemacht eenen groeten bout met eene haeke daeran ende een hoeghe daertoe met twee langhen weren ghebest an de houppranghe, weghende te gaeder 22 pont ... 33 gr[oten].
> (Item first bought in Vinckele for Massijons a brake plank, which cost 3s 4g. Item paid Joos de Soener and Jan Balsan to place the same brake plank in the suitable way, 2s gr. ... Item first made a big pivot with a hook and an eye with two long feathers used at the brake band, with a total weight of 22 lbs ... 33 gr.)
> Rijksarchief Kortrijk, Fonds d'Ennetieres 2975

Halfway through the seventeenth century the town council of Leiden discussed the value of the Flanders *vang*:

> Opte remonstratie off verthooninge bij mr. Aernt van Sgravesandt, fabryck, gedaen, dat verscheyden cooren molens, staende op deser stede vestwallen, haer dienen van eenen Vlaemschen vangh, 't welck naer sijn oordeel niet en was sonder peryckel van brandt, en dat hetselve met het maecken van een lipvangh (sulx als nu heden daechs de meeste molens gebruijcken) soude konen werden voorgekomen, hebben den voornoemde gerechte bij desen verstaen ende geordonneert dat alle de molens op deser stede veste staende off noch te stellen, aen haere molens sullen moeten gebruycken enen lipvangh, ende werden Burgemeesteren geauthoriseert 't selfde te doen versorgen ende uytwercken. Actum de 28 july anno 1648.
> (At a demonstration or exhibition done by Mr Aernt van Sgravesandt, architect, that lots of cornmills, standing on the city walls, are equipped with a Flanders brake, which, according to his judgement, was not free from fire hazard, and that that hazard could be prevented by making a lip brake (as is usual in most of the mills nowadays), the above-mentioned court has ordered that all mills which stand on the walls or are built there in future must be equipped with a lip brake, and mayors will be authorized to take care of the providing. Actum, 28 July 1648.)
> Stroop, *Molenaarstermen en molengeschiedenis*[28]

In 1684 some progressive millers in Leiden equipped their mill with a Flanders brake. The city council judged that the new braking system presented fire hazards and obliged them to use the traditional lip brake.[29]

Improvements in mills always had an empirical character, which accounts for people's suspicious view of innovations. Nowadays it is commonly accepted that the Flanders brake presents a smaller fire hazard. Because of the greater working surface, there is less pressure on the gearwheel, as compared with a lip brake or sprag. The Flanders brake has other advantages too: less wear and smoother handling. The large amount of iron which was needed for the Flanders brake was the reason why that braking system spread only when iron became cheaper and of better quality, something which took place in the fifteenth century.[30] As mentioned already, iron was also used in the block brake. However, the force exerted on the iron fastenings of the Flanders brake is much greater than that between the blocks of a block brake. This might explain why the Flanders brake developed later.

CONCLUSIONS

The conclusions of this paper are as follows. First, the two kinds of brake bands (plank brake and block brake) both already existed at the beginning of the fifteenth century. These two systems do not reveal anything about the kind of fastening. For instance, there were lip brakes with planks and lip brakes with blocks. To find the origin of the Flanders brake, we have to search for other clues.

Second, on the basis of philological clues (*perse*, *pranghe*, *vanghe*) and the mention of iron parts in repairing accounts of braking systems, we conclude that the Flanders brake existed in Flanders only at the end of the fourteenth or even at the beginning of the fifteenth century.

Third, because of its advantages (less wear and a smoother handling) the Flanders brake gradually replaced the lip brake.

For this article we have at our disposal archive data only from Flanders and the southern Netherlands. Investigation of the braking system of mills in archives in other parts of Europe would be interesting and useful. Only then shall we get a clearer view of the evolution of the braking system and the spread of the windmill itself.

Notes and References

1. Rex Wailes, 'Windmills', in *A History of Technology*, vol. 3 (Oxford, 1957) 104.

2. Written statement of Louis Blom (Zwolle, Netherlands, 5 October 1994) to the authors. For Portuguese windmills, see also Anders Jespersen, 'Portuguese Mills: What We Learned during the First International Symposium on Molinology in Cascais in October 1965', *Transactions of the 2. International Symposium on Molinology*, Lyngby, 1977, 74.

3. Rex and Ursula Wailes, 'Notes on the windmills of Öland, Sweden', *Transactions of the Newcomen Society for Study of the History of Engineering and Technology*, 1938–9, XIX: 59; Claude Rivals, *Le Moulin à vent et le meunier dans la société française traditionnelle* (Ivry, 1976), 227; Paul Bauters, 'Een eigen gezicht: Zweedse molens', *Levende Molens*, XVII (9) Sept. 1995: 66. After a journey in Sweden, P. Bauters informed us that Swedish millers did not use the rope for stopping mills.

4. Richard L. Hills, *Power from Wind: A History of Windmill Technology*, (Cambridge, 1994), 31. Hills found his information in M.G. Huard, R. Wailes and H.A. Webster, 'Three Types of Windmills in Southern Brittany', *Transactions of the Newcomen Society* 1949–51, XXVII, 209.
5. Written statement of Louis Blom (Zwolle, Netherlands, 5 October 1994) to the authors.
6. Archives Départementales du Nord (Lille), B 4103, 156 r°.
7. Archives Départementales du Nord, B 4564, 63 v°.
8. Alfred Ronse, *De Windmolens*, (Bruges, 1934), 97.
9. Jan Stroop, *Molenaarstermen en molengeschiedenis*, 2nd ed. (Arnhem, 1979), 17.
10. Anton Sipman, *Molens zoals ze waren en zoals ik hen heb gekend: vang, voering, pal* (Zutphen, 1976), 9.
11. *Marcel Barbier, meunier à Moutier-en-Beauce* (Saint-Denis, 1979), 181.
12. J.L. van der Gouw, *Rekeningen van de domeinen van Putten (1379-1429)*, vol. 1 (The Hague, 1980), 51.
13. Cf. M. van Hoogstraten, *De molens van Zeeland*, 2nd ed. (Middelburg, 1972), *passim*; B.W. Colenbrander, R. Schimmel et al., *Molens in Noord-Holland*, (Amsterdam, 1981), *passim*.
14. Ton Meesters, 'De standerdmolen van Dornum en zijn lipvang', *Nieuwsbrief*, 6, April 1993: 11–15.
15. Written statement of Frans Weemaes (Terneuzen, Netherlands, 9 October 1994) to the authors.
16. Cf. Stroop, *op. cit.* (9), 26.
17. Johann Mattias Beyer, *Theatrum machinarum molarium*, (Dresden, 1767), 75: 'Die Presse wird aus Krümmlingen oder krumm-gewachsenen Hölzern nach der Rundung des Rades zusammen gesetzet: sie dienet um die Mühle zum stillestehen zu bringen, welches geschiehet, wann der Preß-Baum durch das Seil niedergelassen wird, so daß er hernach die Presse um das Kamm-Rad herum anzwinget.' Why did Hermann Gleisberg in *Triebwerke in Getreidemühle* (Dusseldorf, 1970), p. 56, who mentions Beyer's text, replace the first *Presse* by *Preßbaum*, which he put between inverted commas? This confusion is unbelievable and proves that *Presse* and *Preßbaum* are no longer understood by modern molinologists.
18. Luc Goeminne, 'Molentermen te Brugge in de 14e en 15e eeuw', in *Handelingen van het genootschap voor Geschiedenis*, CXXIV (Bruges, 1986), 102 and 'Struktuur en terminologie van de windmolen van Eeklo, Kaprijke en Waarschoot op het einde van de 14e eeuw', in *Appeltjes van het Meetjesland*, XXXVI (Eeklo, 1985), 27, considers *perse* (the brake) and *perseboom* (the lever) to be synonyms, and so makes the same mistake as Gleisberg.
19. Stroop, *op. cit.* (9), 19–25.
20. Anne-Marie Bautier, 'Les plus anciennes mentions de moulins hydrauliques industriels et de moulins à vent', *Bulletin philologique et historique (jusqu'à 1610) du Comité des travaux historiques et scientifiques*, II (Paris, 1961), 567–626.
21. In a desperate attempt to prove that the windmill was an English invention, Kealey studied the oldest English records of mills, but even his theories encountered a lot of resistance from, among others, Holt. Edward J. Kealey, *Harvesting the Air: Windmill Pioneers in Twelfth-Century England* (Bury St Edmunds, 1987); Richard Holt, *The Mills of Medieval England* (Oxford, 1988).
22. Stroop, *op. cit.* (9), 27; H.G. Hamaker, *De rekeningen der grafelijkheid van Zeeland onder het Henegouwse huis*, I (Utrecht, 1879), 138, 158.
23. We read the word *ghevanghene* in an account of Wervicq-Sud (Nord, France): camwiel met 52 cammen met eenen cruuse duer den asse gaende met tween stuenen ende dobbel ghevlecht ende *de ghevanghene*, but it is probably a wrong interpretation made by the copywriter; in the estimation of the same mill made in 1502, we read: 't camwiel met 52 cammen, een cruus duer den asse gaende met tween stuenen, dobbel ghevlecht metgaders *de vanghe*. Archives Départementales du Nord (Lille), B 19671 (years 1496 and 1502).
24. On the problem of technical nomenclature see J. Needham, *Science and Civilisation in China*, Vol. V, Part 7, 'Military Technology – The Gunpowder Epic' (Cambridge, 1986), esp. pp. 1–12; G. Hollister-Short, 'The Vocabulary of Technology', *History of Technology*, Vol. 2 (London, 1977), pp. 125–55, and 'Harmless Drudgery: Useful Work', *Technology and Technical Sciences in History*, Proc. ICOHTEC Symposium (Dresden, 1986), 123–6, which cites recent French work on this subject.
25. Oral statement of J. den Besten (Loenen aan de Vecht, Netherlands, 1994).

26. Some accounts from the sixteenth century mention a *langue* in connection with a bridge tree for the millstones. This element has about the same function as the *langue* of the brake system, to connect two levers. Cf. Archives hospitalières de Lille, I 4301, 48 r° Lille (1527).

27. Cf. The mill of the *Encyclopédie* of Diderot en d'Alembert. Cf. also *Marcel Barbier* (note 11) p. 181: 'Dans l'ancien temps, le levier du frein était toujours au premier étage. Après, les meuniers les [*sic*] ont montés au deuxième, j'sais pas pourquoie?' (Formerly the brake lever was always at the first floor. Later, millers placed it at the second floor, I don't know why).

28. Quoted in Stroop, *op. cit.*(9), 30.

29. From this record from Leiden Stroop concludes the exact opposite. The new lip brake should have replaced the older Flanders brake!

30. Even in accounts from the end of the fifteenth century we read that the iron spindle of the mill often broke.

Nuremberg as Epicentre of Invention and Innovation towards the End of the Middle Ages

WOLFGANG VON STROMER

From the beginning of the fourteenth century the imperial city of Nuremberg, along with Cologne, developed into the most important commercial city of the Holy Roman Empire, and between 1335 and 1340 overtook Regensburg in importance. It became the major central European exporter of goods. Its network of commercial contacts was densely woven, covered the whole of central Europe, and stretched beyond that across the whole continent from the North Sea to the Mediterranean. All this was an achievement of its citizens and an outcome of their co-operation with the rulers of Germany alone, since its geographical setting was scarcely propitious for such a role. The success of the Nurembergers rested upon an unwonted number and variety of innovations, intensified by the interplay between them. These innovations were partly of an intellectual-organizational nature and partly consisted in technical inventions in a variety of fields. Some have long been the object of attention and admiration while others have become evident only as a result of recent systematic research. The space available here permits little more than a listing of them.

ORGANIZATIONAL AND INTELLECTUAL INNOVATIONS

(i) Nuremberg's great success in wholesale and foreign trade originated in its unique system of commercial privileges in 70 different centres and in several central European lands. Nearly a century ago the economic historian K. Th. von Imama-Sternegg described the Charter of Privilege of 12 September 1332, in which the Emperor, Ludwig the Bavarian, confirmed these 70 privileges in the kingdom of Arelat (Arles) and on the Danube, as 'the greatest achievement in the field of economic history of the Middle Ages'. Thanks to the work of the Swiss historian Hektor Ammann, we know also that these commercial privileges had their origin in a major programme of imperial economic strategy inaugurated by the Hohenstaufen Emperors Friedrich I, Friedrich II and Philipp of Swabia. Other cities in addition to Nuremberg received privileges, including the six other cities (*Kammerstädte*) of the Imperial Chamber: Metz, Arles in Provence, Cambrai, Frankfurt, St Trond and Aachen.[1] The fall of the Hohenstaufen (1254) brought an end to this important concept. In the spring of 1264, however, the merchants of Mainz and Nuremberg renewed it among themselves, and agreed to mutual freedom from customs. From this point onwards the Nurembergers worked at establishing systematically this method of setting up a system of mutual freedom from customs, which was eventually confirmed by Ludwig the Bavarian in 1332. The principle had, however, already been recognized as one of Imperial law when the Emperor Heinrich VII granted them a privilege in Pisa in June 1313.[2]

In the course of the fourteenth century the great commercial firms of Nuremberg were able to obtain additional privileges for themselves and for their fellow citizens which were valid for whole countries: Brabant 1311, Bohemia and Moravia 1335, Brandenberg 1348, Hungary with all the lands belonging to the Crown of St Stephan 1357, Flanders 1362, Poland and Galicia 1365, Milan, Lombardy and Genoa 1346–98, and the kingdoms of Spain 1415 (Figure 1). In order to obtain such charters of privilege the Nurembergers opportunistically exploited, before all else, economic and political tensions. For example, they received their important Flemish privilege as thanks for breeching the great blockade of Flanders of 1358–60 by the Hanse.[3] They achieved the removal of discriminatory measures against German traders in Venice by means of an economic war which they conducted with the help of Ludwig IV and Karl IV from 1340 to 1348 and in 1358. Venetian merchants were subject, as foreigners in the Empire, to a 'road directive' by which they were forced to travel overland solely by way of the *caminus Norimberghe* and the *caminus Basle* if they wished to reach the great markets of north-west Europe, above all Flanders. The overland route was the only practicable one for goods such as silks and spices, both extremely susceptible to water damage. These two roads were the only routes available for the express couriers engaged in spice and currency speculation. Co-operation between Nuremberg and Basle meant that pressure could be successfully exercised, and in the event of hostilities breaking out, all connection could be cut off. Such methods brought Nuremberg a favourable compromise in 1348 which gave her completely reciprocal rights with the Serene Republic, and in 1358 even a long-term exclusion of Venetian merchants from trade with Germany in the lands lying between the Rhine and the mountain frontier in the east.[4] Thanks to the advice and help of prominent Nuremberg merchants, the Emperor Sigismund was able to wage war against Venice in 1412–14, and in 1417–33 in Dalmatia and the Imperial lands. On the Terraferma (Venetian mainland) hostilities took the form of a successful economic war, one which led directly to the establishing of a Continental blockade in the manner of the later Napoleonic blockade. As thanks and reward the Emperor granted the Nuremberg merchants in return commercial privileges in Lombardy, Savoy, Genoa and its colonies, and the Spanish

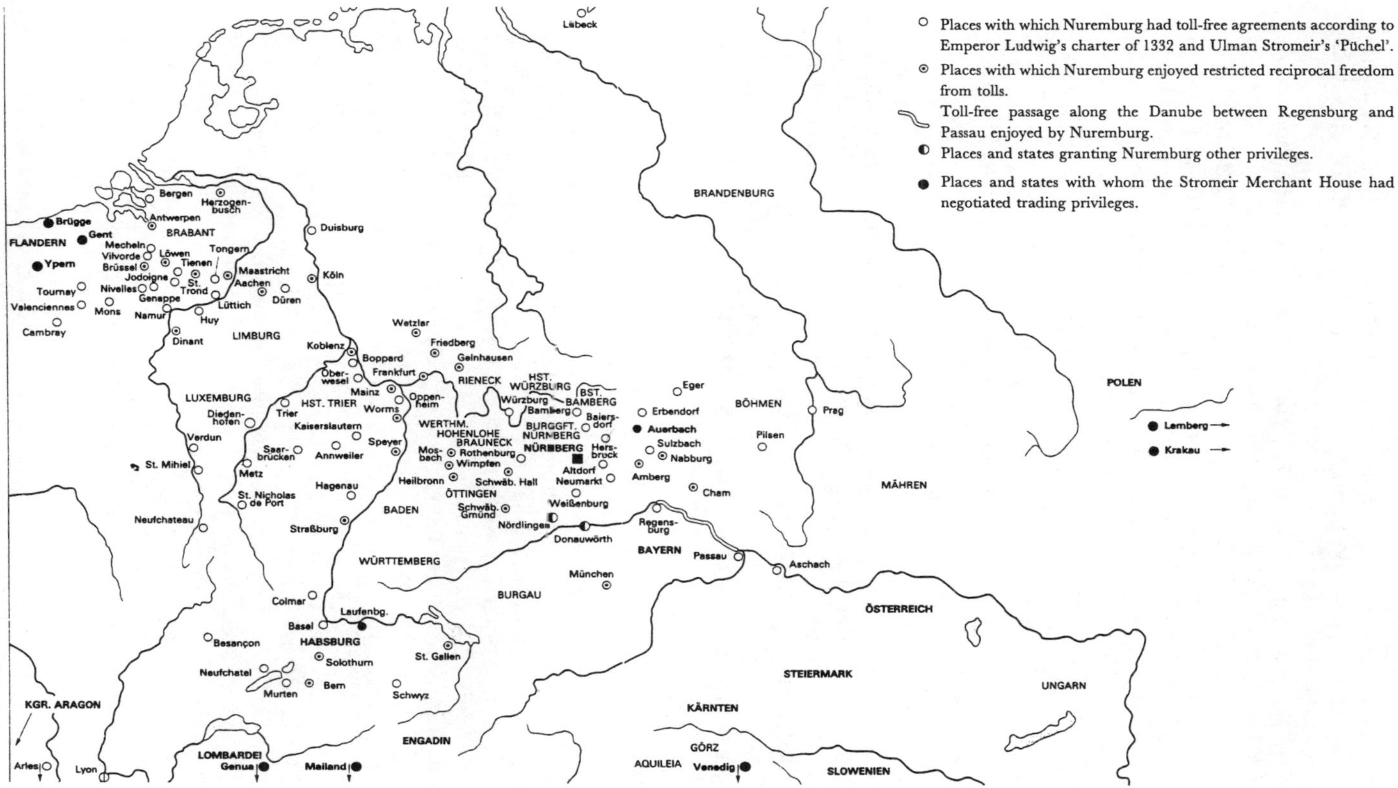

Figure 1 Nuremburg's trading agreements of the fourteenth and fifteenth centuries.

kingdoms, and opened up for them new commercial routes from the mouth of the Danube to the Bosporus and to the still extremely important emporia of the Black Sea.[5]

After 1377, six major Nuremberg commercial firms were able to gain sole control of six of the 56 departments (*Kammern*) in the Fondaco dei Tedeschi on the Grand Canal in Venice, and maintain their hold over them for several generations. This gave them special advantages in their trade with Venice and the Levant. Before this, from 1337 they were able to displace the merchants of Regensburg as chairmen of the *tavola* of the south Germans, and from that point onwards they held the key positions in the Fondaco, with Nuremberg merchants holding the majority of the German consular seats.[6]

(ii) This novel and singular way of using commerce for political ends was based on unprecedented forms of organization and structure deployed by Nuremberg firms, and extended also to their methods of production. The unusual extent to which the division of labour was pushed, with manufacture subdivided into more than 200 different crafts, was co-ordinated with the *Verlag* system of production and distribution (*Verlagssystem*). In such a system the great firms, as *Verleger* (co-ordinators), controlled each step of production, for example with knives or needles. Rationalization on such a scale brought about a considerable improvement in the quantity and quality of standardized mass-produced goods, which thereby attained an almost unrivalled export position in foreign fairs and entrepôts. Adam Smith in his *Wealth of Nations* of 1776 regarded the efficient division of labour as one of the chief causes of a country's prosperity. His illustration, drawn from pin manufacture, had an earlier exemplar in the needles of Nuremberg. Although certainly existing as a form of organization for some time previously, the concept of *Verlag* and *verlaging* is first mentioned in 1340 in an ordinance of one of the Nuremberg crafts. Not surprisingly, it was one of the metal crafts.[7] After 7 January 1387 Nuremberg firms controlled at least 19 out of the 82 ironworks (*Eisenhämmer*) belonging to the Großen Oberpfälzer Hammereinigung (the Great Oberpfälzer Ironworks Union), with 21 smelting furnaces also operating under the *Verlag* system in the Nuremberg–Upper Palatinate mining area (Figure 2). The Ironworks Union (Hammereinigung) itself provided a cartel for the 82 hammer works and their 75 entrepreneurs that permitted them to control supplies to five great areas of scarcity in the iron industry. This is the first cartel known to us in the medieval economy, and was extremely successful in its first projected operational period of only four years. Later, as it was continually renewed down to 1616, its sustained restrictions on production had long-term consequences for innovation, and crippled the whole mining area of the Upper Palatinate.[8]

Nuremberg businessmen employed the instrument of the *Verlag* in all other branches of industry. Ulman Stromeir handed over the first German paper-mill, which he had established between 1394 and 1398 as a going concern with a *Verlag* agreement (the first which exists word for word in

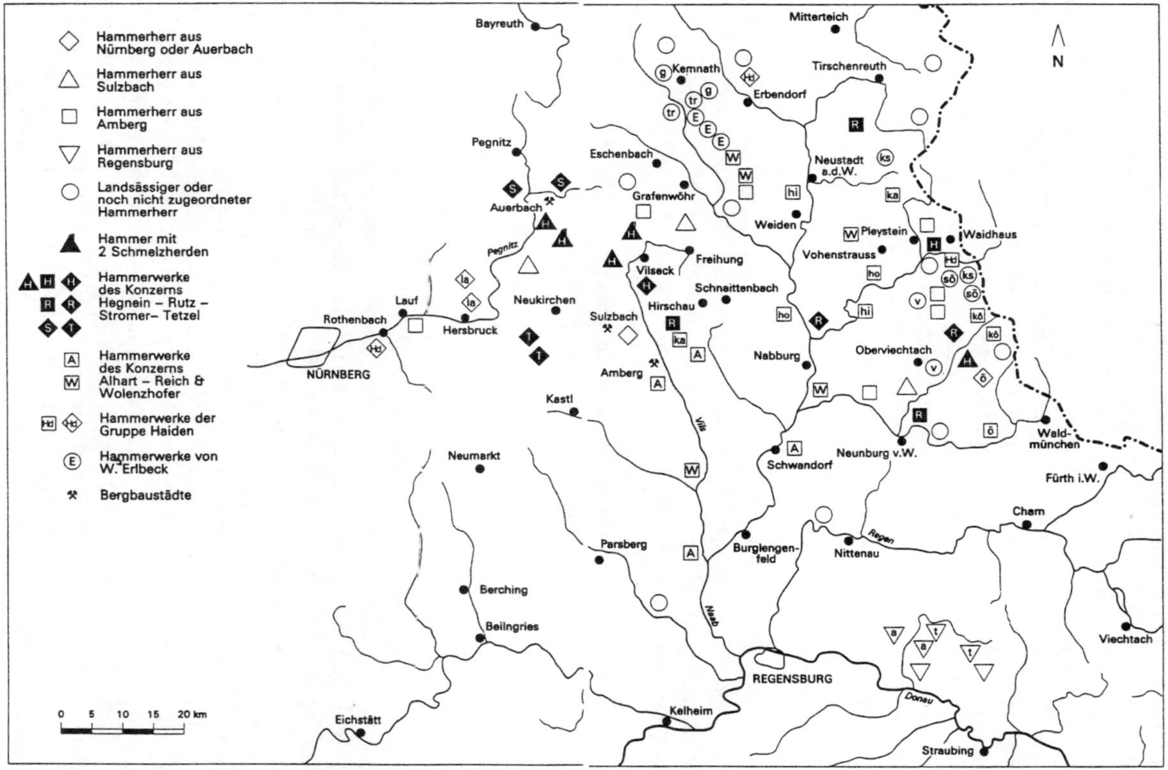

Figure 2 The Oberpfälzer Ironworks Union of 1387.

the original), to the paper producer Jörg Tirmann, whom he had initially hired for 10 years as a professional paper-maker in August 1390.[9]

Nuremberg entrepreneurs were able to employ the instrument of the *Verlag* to develop whole areas of textile production; for example, from 1396 in Ries, Nördlingen, Bopfingen and Lauingen, from 1411 in the Vogtland of Upper Franconia with Hof and Münchberg, and in Upper Hungary (today Slovakia) with Kaschau (today Kosice) and Bartfeld (today Bardejov) as centres. After 1480 a major area for linen production came into existence in the eastern part of central Europe with 37 smaller towns as centres in Saxony, in Lausitz, Lower Silesia and in the Sudetenland, which transcended natural and political borders, and even survived the Thirty Years War. The Nuremberg firm of Peller & Viatis was able to fulfil the needs of slave traders for three-spun material for the whole of the New World (blue jeans *from* Gênes, Genoa).[10]

The ever-growing demand for capital on the part of these firms was combined with a high liability risk for the small investors such as the apprentices, who, although they partook of only a modest part in the profits, carried full liability for the debts of the whole firm. The Nurembergers managed, however, thanks to a charter of privilege of the Emperor Friedrich III of 1464, so to arrange matters that only the heads of firms were liable with their full property and eventually could be thrown into debtors' prison, whereas the small investors were liable only for the sums they had invested. As well as the usual business firm, an additional form of organization and kind of law had thereby come into existence: the *Kommanditgesellschaft* or, in Italian, the *Commenda* (a limited-liability company).[11]

(iii) The pioneers in Germany of the essentials of mercantile bookkeeping were Nuremberg firms expert in this field. The first German example of bookkeeping, the *Handlungsbuch der Holzschuher*, drawn up between 1304 and 1307, contains a systematic list of debtors drawn up according to their social status – nobles, clergy, townsfolk – and thus, according to their status under the law. The account books and bills of the firms Kress and Paumgartner from 1398 and Marquart Mendels's 'Business Book' of 1425–38 are written in Venetian style – exactly as learned on the spot – with debit and credit entries (Figure 3). The Hans Praun firm, which dealt in books as far away as Bologna, and also engaged in silk manufacture there, was the first northern European firm to adopt double-entry bookkeeping (after 1479). This improved bookkeeping is first observable in Italy after 1428 in the business records of the Borromei of Milan. Double-entry bookkeeping was employed by the Nuremberg-Milan firm Köhler & Kress & Saronno in 1496–1511.[12]

(iv) Around 1395 Ulman Stromeir composed his 'Püchel von meim geslecht und von abenteur'(Book about my family and business ventures).

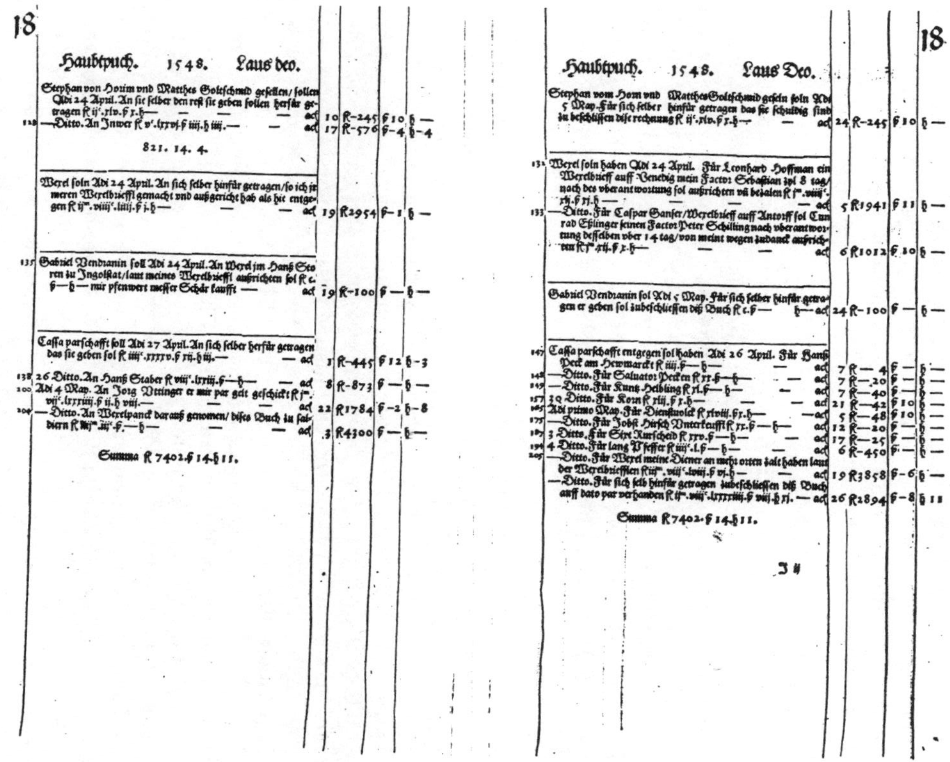

Figure 3 Sample pages of double-entry bookkeeping. From the first German textbook of 1549 by Johann Petreius.

With its account of the establishment of the first paper-mill in central Europe in Nuremberg in 1390, it also served as a record of this undertaking, and with its chapter on weights and usages served as a description of commercial practice right across Europe from Barcelona to Azov on the Don. It is the first German *practica della mercatura*.

From 1383 the correspondence of various firms shows that a new product, fustian, was being distributed, employing samples; that bills of exchange were being employed for the transfer of money between Nuremberg, Milan, Genoa and Venice; and that arbitrage (currency speculation) was going on. The next German example of the use of samples in trade comes from 1454–55, when the *vir mirabilis* Gutenberg was advertising for customers for his bible by exhibiting pages (Quinternions) at the Frankfurt autumn book fair and by sending specimen pages to important people such as the Emperor and the cardinals, offering them his bible for sale.[13]

Advanced bookkeeping offered the entrepreneur better information about the internal running of his firm, and thereby reliable data on which to base future development. Most important was the opportunity which this gave him to optimize his credit potential and thus create money on paper in the same way as he did in trading with bills of exchange. Insufficiency of gold and silver production due to technical problems in the mines then in active exploitation meant that there was a lack of precious metal coinage throughout the Middle Ages and early modern period, and this was one of the chief causes of sustained economic depression.[14]

(v) Ascertainably after 1365, the money-changers of Nuremberg contributed to the creation of giro money in that they developed almost all the functions of modern banks. The account books of money-changers that have survived acquaint us pretty well – thanks to the exact credit and debit entries – with how this giro system functioned. Only Frankfurt from about 1400, with its famous fair, offers equivalent records for late medieval Germany. The prototypes for the exchange offices of these two economic centres were the exhanges of Venice, Milan, Genoa and especially Bruges, where German firms had operated. In 1618, however, Nuremberg merchants established a *banco pubblico* parallel to the establishment of a public bank shortly before in Hamburg. This bank possessed its own private artificial currency, accepted privately and internationally as the *mark banco*. Thanks to the corresponding investment of the bank members in silver bars, this currency was accepted at face value.[15]

In Nuremberg in 1482 were printed the first handbooks and books of instruction relating to business mathematics, initially tables; that is, basic arithmetic. These were followed by more advanced handbooks on bookkeeping, the rules and law of exchange, and exchange usages.[16]

TECHNOLOGICAL INVENTIONS AND INNOVATIONS

(i) Nuremberg's reputation in craft skills and in the field of technical invention, especially for metalwork and fine mechanics, was far greater than it was for the intellectual and economic innovations mentioned above. Between these two areas, however, great successes in cartography, surveying and technical drawing, geography and astronomy, combined with navigational achievements, ensured that Nuremberg played a major role. The Nuremberg 'Schreibmeister', Johann Neudörfer the Elder (1497–1563), after numerous publications about his own art, in 1547 published *Nachrichten von den vornehmsten Künstlern und Werkleuten, so innerhalb hundert Jahren in Nürnberg gelebt haben* (Information about the most distinguished artists and craftsmen who have lived in Nuremberg in the past hundred years). In 1730, Johann Gabriel Doppelmayr followed in Neudörfer's footsteps with his *Historischen Nachricht von den Nürnberger Mathematicis und Künstlern* (Historical notices of Nuremberg mathematicians and artists). Both are invaluable sources for cultural and artistic history and for the history of technology.[17] Particularly important here is the *Ephemerides* of Regiomontanus of 1474. The work, in 896 printed pages and with around 300,000 numbers and symbols organized in tables, sets out the position of the stars, mostly determined from Regiomontanus's personal observations, projected for the years 1475–1531. It was a work that opened the way for the navigational achievements of the age of discoveries since it was now possible, for the first time, to determine one's position both on the open ocean and in relation to unknown coastlines. Columbus, Amerigo Vespucci and Vasco da Gama all used Regiomontanus's work.[18] Martin Behaim's globe of 1492 represented the earth as a ball for the first and last time before the great discoveries. However, Cardinal Nicholas of Cusa had been able to obtain a globe of the heavens with the traditional representation of the stars as early as 1444 from Nikolaus von Heibech, '*orloymeister bei dem Stromeyr*' (the Stromeirs' clockmaker), together with a novel surveying instrument, the 'Torquetum'. Although scholars have scarcely noticed either of them, both still survive in the keeping of the Cusanus Foundation in Berncastel-Kues.[19] It was a century later in Nuremberg that Nicholas Copernicus's *De revolutionibus orbium coelestium* first appeared in print. It led to a complete transformation of the way in which the world was perceived.[20]

Regiomontanus announced the printing of a *Mappa Mundi* in 1474 which unfortunately has never been discovered. About 1500, the best cartographer of the day, Erhard Etzlaub of Nuremberg, produced his map of the route to Rome which, with its system of mile-points, was more accurate than modern maps such as the Shell atlas.[21] Between 1525 and 1530 Albrecht Dürer followed this up with his works on proportion, surveying and the fortification of towns and fortresses. Masterpieces of cartography produced in Nuremberg include Paul Pfinzing the Elder's *Atlas* of 1594 and his textbook *Methodus Geometrica* of 1598.[22]

Between 1582 and 1614 the Ratsbaumeister, inventor and architect Wolf Jacob Stromer, produced twelve volumes of a textbook on the art of building and urban planning. The Fleischbrücke which he constructed between 1596 and 1598 was regarded by contemporaries and professionals down to the nineteenth century as a superlative example of bridge-building technology. With its foundations built in marshy ground, it spanned 30 m of water in a single bow, as did the Rialto Bridge constructed in 1591. The difference is that the Rialto Bridge is a quarter circle, whereas the segmental arch of the Fleischbrücke comprised a fifth of a circle. Its correspondingly gentle gradient made it passable for heavy vehicles.[23]

(ii) Nuremberg entrepreneurs secured the *nervus rerum* – silver – and the raw materials for their metal industry, whether for weapons, tools or fine mechanics, by means of technological improvements in the mining of precious metals, copper and iron ore, and in the refining of these metals. In the interests of progress, both in mining and refining, they were prepared to invest and engage in sustained experiment for years at a time, and even to bring in Jewish experts.

The exhaustion of the ore deposits of central Europe had been a fact of life since the thirteenth century in respect of precious metals, and since the fourteenth century as far as the deep mining of copper and iron was concerned. The chief problem in the way of winning extra metal was that of keeping mine workings clear of water. Many of the best reserves lay beneath groundwater level and were in danger of being lost by flooding. Some which could no longer be kept pumped dry, and thus were overwhelmed with water, were known to contain rich, but unreachable, deposits of ore. Mining concerns in Nuremberg were constantly preoccupied with developing new kinds of water engines – *Wasserkünste* – which, with the help of experts, including the above-mentioned Jewish *Künstiger* (artificers), they constantly sought to adapt, refine and modify according to particular circumstances. Whole genealogies of expertise and know-how were developed by firms which guarded their knowledge with an almost modern zeal for secrecy. The experts and their firms – and there was competition in the development of pumps between different firms – took part of their pay from the profits from ores won by their expertise, or as shares in the very mining ventures themselves.[24]

(iii) Washing the ores, working pumps, smelting and smithing or casting metals demanded great quantities of various kinds of energy. Fire, water and manpower were employed. The growing cities and their crafts and the mining industry all exploited the neighbouring forests to supply the essential sources of heat energy, wood and charcoal. The woods were thus annihilated. From the end of the fourteenth century cities like Nuremberg and Frankfurt, and mining towns like those of the Upper Palatinate which lived by the export of their industrial goods, sought in vain – through rigorous control, limitation of usage and closure of industries which consumed large quantities of fuel – to prevent the further destruction of the forests. But no ready-made solution was at hand. Nuremberg experimented

Figure 4 Peter Stromeir, 'father of forestry'. Copied from a posthumous portrait, oil on beech, of *c.* 1430.

from 1309, certainly with many setbacks and disappointments, until the mining entrepreneur and merchant Peter Stromeir (d. 1388; Figure 4) was at Easter 1368 first able to plant a pine forest successfully in the Imperial domains near Lichtenhof – 'davon nu gross vil weld kumen sein' (from which great woods have now grown) as his brother Ulman noted a generation later in his '*Püchel*'. The method – harvesting the seed from not quite ripe pine cones of well-chosen 'seed trees' in autumn, the artificial ripening of the seeds by means of a special heat treatment, sowing the seed the next spring in an especially deeply ploughed area in the forest – required trial, error and accumulated experience. This Tannensäer (fir seeding) technique remained for several centuries a closely guarded secret of the Nurembergers, the city making gifts of sacks of seed to friendly rulers and allied cities. The principles of this technique are still employed today in forestry in the temperate zones of the most heavily industrialized parts of the world.[25]

A second kind of energy shortage resulted from the plague of 1348 in which approximately one-third of the population of Germany perished, leading to a severe shortage of manpower. An attempt was made at once to

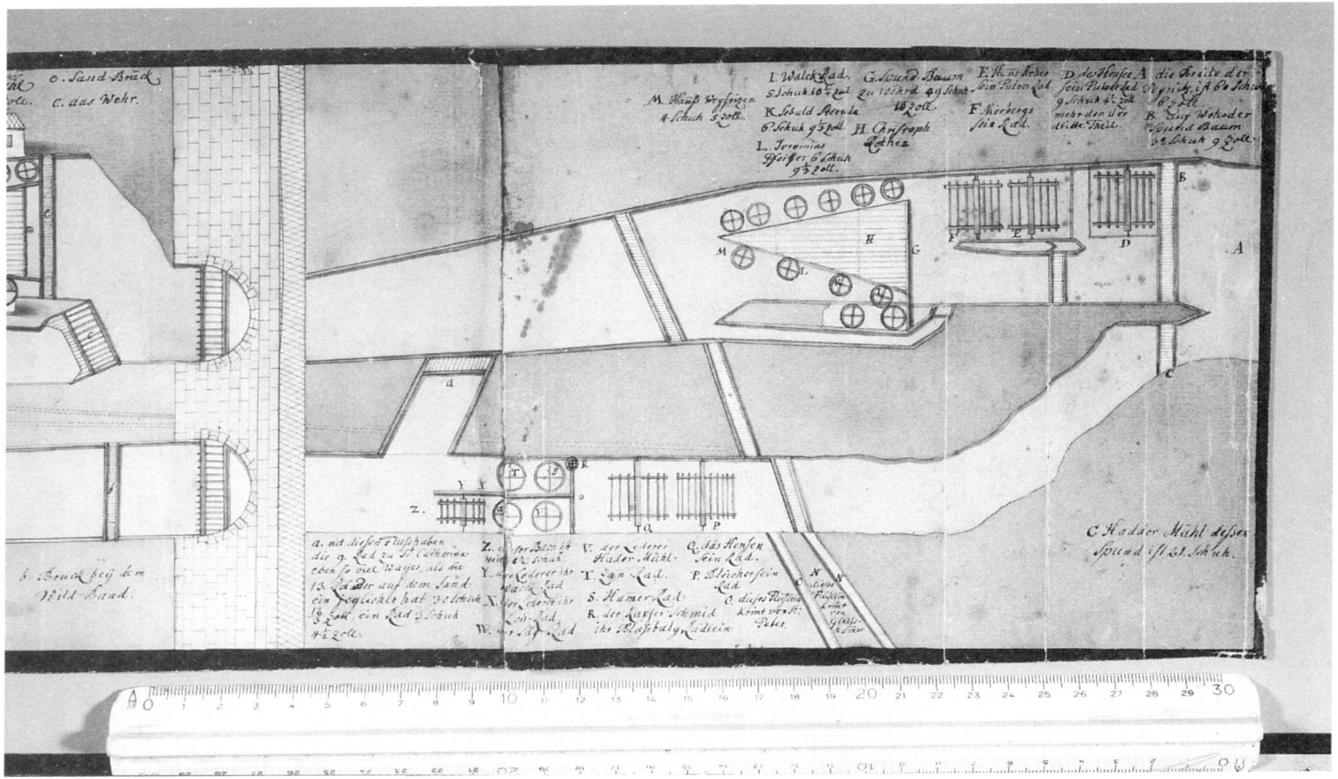

Figure 5 A variant of a sketch of 1601 showing the names, functions and positions of certain of Nuremburg's 131 waterwheels.

replace human labour with water power. New machines were employed in the iron industry of the Upper Palatinate to drive the mills with cam-actuated hammers. The *fabricae pedales*, used in the district since time immemorial, and whose huge bellows had previously been worked by the feet of labourers, were now driven by water power. Thus were the refineries and forges now supplied with forced currents of air. Suitable mill locations in the neighbourhood of cities had long been mostly occupied by grain mills but now, in addition, hammer mills as well as saw mills, fulling mills and stamping mills (for comminuting ore and gunpowder and quartz for glass), and soon also paper and wire-drawing mills, were competing for sites on the rivers. Wherever it was at all practicable, rainwater and water from streams was stored by creating dams and millponds, which since the Middle Ages have dominated the landscape of proto-industrial areas. On two former locations of the Nuremberg-Upper Palatinate industrial area, namely, on the Pegnitz in the towns of Nuremberg and Lauf, water power was sufficient to drive 131 and 44 water wheels respectively (Figure 5). Where the river split into several channels, as it did outside the walls of Nuremberg, the flow was measured with great accuracy down to one-quarter of an inch (or near enough 1 cm) where it was divided between the weirs. This we know from the example of Ulman Stromeir's paper-mill in 1397. This remained the case until modern times.[26]

(iv) The search for scarce metals for minting into coin could not be satisfied even after the discovery of new ore deposits and the exploitation of older mines to greater depths, with improved pumping engines. Money changers and coiners specialized therefore in taking out of circulation coins that were heavier than usual or had a greater precious metal content – ascertained respectively by the use of delicate scales and touch needles. Such coins, together with old money and foreign pieces and broken bits of silver jewellery and ornaments, had their silver and copper content extracted from them by means of an exacting metallurgical process called *prennen und seigern* (burning and liquating). The silver was released in the smelting process by the addition of lead, which latter was then oxidized. The process required huge quantities of lead (whose arsenic content poisoned the surroundings of the smelting ovens) as well as huge quantities of charcoal. The procedure was forbidden since it reduced the quality of coins in circulation and was regarded as a kind of forgery. It demonstrated the fact, however, that using the same process one could thereby transform the nature of the brittle *Schwarzkupfer* – that is, black copper – won from the ores of the mines of central Europe. As a result, a malleable – that is, easily forged and drawn – refined copper (*Garkupfer*) became available. After 1367 German experts – presumably from Nuremberg – worked in the Hungarian mining zones on the development of this technology. Between 1390 and 1395 they had succeeded sufficiently to be able to outbid the Florentine mining company of the Medici and Venice with its copper-smelting works in the ghetto to attain a dominant trading position in copper on the world market.

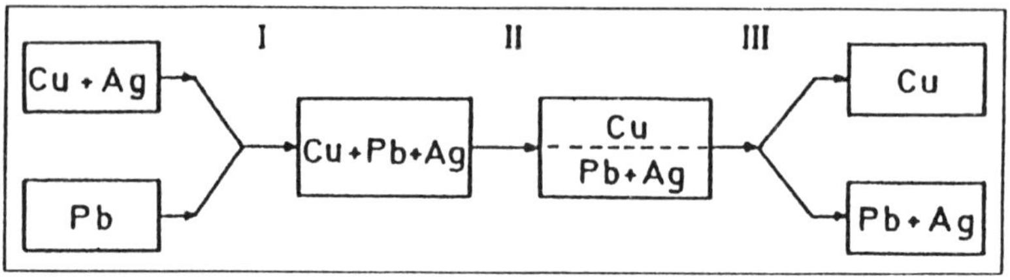

Figure 6 The principle of copper separation: 'ars conflatoria separantia argentum a cupro cum plumbo'. *Source*: Lothar Suhling, *Der Seigerhüttenprozeß* . . . (Stuttgart, 1976), 18.

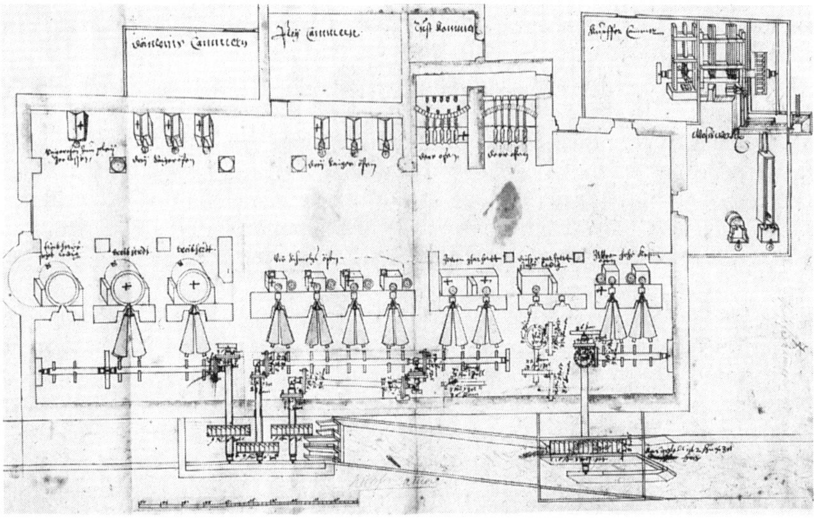

Figure 7 Drawing of a Saiger furnace of *c.* 1580–1600, probably the Hütte Leutenberg.

Between 1419 and 1453 the smelting furnaces of the city of Nuremberg (located in the former city moat near the Findelgasse) grew into a huge undertaking, as they were now used as Saiger furnaces. They were able to separate ever greater quantities of silver from the black copper won from the ores of the central European mines, which ores contained anything between 0.6 and 1.8 per cent silver (Figure 6). In spite of the enormous cost of charcoal and lead, it appears to have been a profitable process, so that within a decade, between 1450 and 1460, six Saiger furnaces were operating in and around Nuremberg. According to the standards of the day these were major industrial and capitalist undertakings in which highly qualified and highly paid metallurgists and skilled workers employed multi-phase processes in separating and refining metals with the highest precision and economy. This is a very different picture from the still widespread and apparently ineradicable notion of a technologically and economically stagnating Middle Ages (Figure 7).

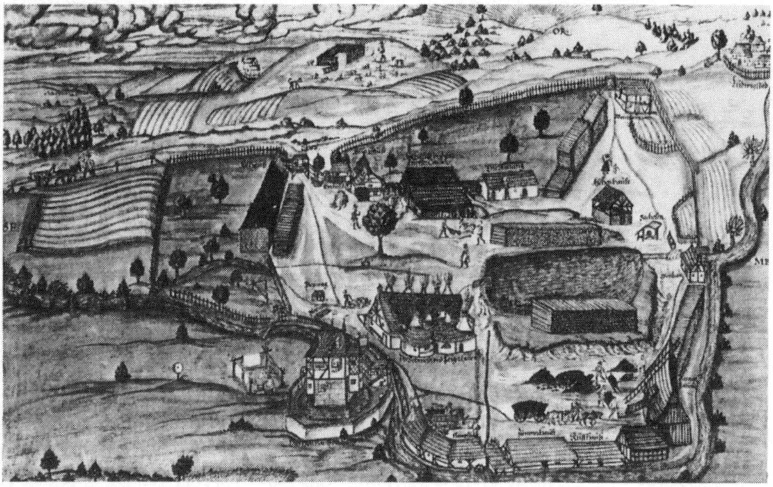

Figure 8 The Ludwigstadt Saiger furnace (constructed 1486) from Paulus Pfinzing's *Atlas* of 1594.

In addition to the production of silver in cupels, the by-product, *Garkupfer,* the refined copper, was also a new and highly desirable material. Unlike the previously available black copper, it was easily wrought at the forge and was highly ductile. These characteristics were still further improved by the addition of calamine – zinc carbonate – in order to produce an alloy, brass. This resulted also in the melting-point of the metal being reduced. As a consequence, bells, fountains, pipes and numerous everyday articles could be produced more cheaply. Cheaper pans and kettles could be made, metal sheets formed and wire of every calibre could be drawn. I shall return to this point.

The numerous advantages of the Saiger furnaces for the entrepreneurs, for the many crafts and for the economy of the Imperial city overall were taken carefully into consideration by the ruling inner council in considering the future of the industry. A number of the entrepreneurs themselves, or senior members of their firms, sat on this council. The quantity of wood and charcoal burned in the smelters was so enormous that the price of these goods was rising far above what ordinary people, craftsmen and manufacturers could afford. The remaining forests available were under such threat that the very sustenance of the city as a whole *vis-à-vis* fuel supplies was placed under severe threat. The common interests of the citizens were regarded in this situation as having priority over private interests. The result was that so many restrictions were imposed upon the smelting furnaces that most of them were forced to go out of business within a few years. The enormous investment in building them – each Saiger furnace consisted of at least six to eight smelters with their associated waterwheels for working the bellows – had to be totally written off. The industry as a whole was transferred bodily to the forests of

Thuringia, where apparently inexhaustible quantities of wood and unexploited watermill sites were still available. The transport routes to the copper mines of the Harz and the lead mines in Lesser Poland were also thereby shortened. In an enormous wave of innovation a heavy industry with ten Saiger furnaces was built up in Thuringia between 1461 and 1505 (Figure 8). The ruling families of Nuremberg had carried through a unique economic and social achievement.[27]

(v) There was an enormous demand for the standardized goods produced by the Nuremberg craftsmen and proto-factories, whether for arms and high-precision tools or for wire and wire goods. The *merze del Fontego* (warehouse goods), such as knives, scissors, candlesticks, candelabra, horns and needles which Nuremberg delivered in huge quantities to Venice, was the freight most in demand for the galleys sailing to the Levant. Goods in similar volume and of like variety found a ready market throughout the continent, at the great annual fairs and in the ports active in the export trade. The special quality of Nuremberg scissors in the fourteenth and fifteenth centuries has been discussed by me recently elsewhere.[28]

The enormous demand for Nuremberg wire and wire products could not be satisfied with methods employing the usual manual technology. To produce wire as thick as a pencil lead from wire as thick as a pencil (previously worked up under water-powered hammers) using any of the metals available – iron, copper or bronze – meant that it had to be drawn manually fifteen times through a steel plate. This necessitated enormous drawing power, which in the initial stages of production required the wire to be wound round a drum whose capstan-like bars were operated painfully by a number of workers. About 1360 the Nurembergers invented the *Schocke*, a kind of swing attached to a bell crank mechanism with a long arm on which a number of *Schockenzieher* (pullers) could pull simultaneously, and a short 'power' arm to which a pair of pliers was joined by a short chain. This enabled a further piece of wire to be pulled through the steel drawplate with each backward pull. The pliers had to be reapplied after every movement, and the bite of the pliers naturally spoiled the form of the wire (Figure 9). This was tolerable for needles and similar wire products for which only short lengths of metal were necessary, such as the rings of chain-mail, but growing demand required an increase in production which could not be satisfied however many workers were employed. After 1399 systematic attempts were made to mechanize wire production using water power. There were many complex problems to be resolved. The metal feedstock had to be homogenous, smooth and free of slag particles so that the wire would not break or become misformed and thereby unusable. Attempts to produce such metal feedstock were taking place at the very time of the experiments in the smelting process mentioned above. It took some 15 years of costly experimentation before a method was developed of adding calamine (zinc carbonate) to produce a forgeable and ductile brass suitable for drawing and equally to solve problems associated with the machinery.

Figure 9 Manually operated wire-drawing apparatus, Biringuccio, *De la Pirotechnica* (Venice, 1540).

The solution was to construct pliers that automatically closed round the metal wire when the assembly was pulled, and automatically opened on the return stroke. The rhythm of this work was predetermined by the circular movement of the waterwheel according to its diameter and according to the flow of water which thereby powered the machinery, consisting of a crankshaft with a connecting rod upon which the automated pliers were mounted. The full development of a completely automated mechanism was not quite completed within the 15 years. A skilled worker was still needed to fit the pliers initially round the metal so that it could be drawn through the draw-plate. Such a worker was called the *Schockenzieher* (the swing puller), since he sat on a swing (Figure 10). This permitted him to move more easily with the rhythm of the machinery. Since this was centuries before Galileo and Huygens had enunciated the laws of the pendulum, setting up a swing needed a good deal of trial and error before it functioned properly. For the whole period of these experiments the mill *am Sand* on the northern arm of the river Pegnitz brought in no profit whatsoever, and the city council granted it freedom from all dues until, in 1415, the wire-mill really began to function properly. Within the next two generations 30 wire-mills were opened in and around Nuremberg, and the expansion of wire production led to the development of a new industrial district. In spite of a number of attempts by outsiders to obtain the secrets of the process or to lure away skilled workers, the Nurembergers were able to keep the semi-automatic wire-drawing process secret for three generations, and to maintain their monopoly of wire products of all kinds not only in Europe but also in the rest of the Mediterranean world. When they finally lost their monopoly, they did not give up but managed, in the sixteenth century, to develop a fully automatic wire-drawing mill (Figure 11), the technology of which was superseded only in the nineteenth century.[29]

Figure 10 Semi-automatic wire-drawing machinery, Biringuccio, *De la Pirotechnica* (Venice, 1540).

(vi) Numerous everyday objects were fabricated from the wire made in Nuremberg, produced by employing an unusually highly-developed system of division of labour. These objects included needles, pins, safety pins, hooks and eyes, delicate chains, fishing hooks, birdcages and fish- and mousetraps, and also sieves. These last were important not only for household use but also as machine parts for industry, as for example for use in grain-mills (for bolting flour) and for use in pulverization processes, as in the preparation of ores, dyestuffs and gunpowder. For the paper-mills the sieve format was identical with that of today's paper: Imperial (DIN A4!). The finer and more even the wire, the finer and more even the mesh of the net, and the better, therefore, the quality of paper for writing, printing and as material for prints and graphics. Although it was possible to write almost as well on the earliest handmade paper (*Büttenpapier*) as on parchment, it was unsuitable for the reproduction of drawings (except for relatively rough woodcuts or for playing cards).

A smoother and more even quality of paper was needed for the pictures of miracles and saints and for the block books; that is, books printed from wooden blocks on to which the text had been incised. Finer sieves were thus required, and in addition the paper had to be polished to produce a fine surface. Polishing as well as pressing the still moist mat of paper between felt-covered surfaces (*Gautschfilze*) was first developed in western paper-making in the rag-mills (*Hadermühle*) of Nuremberg. Here large pieces of coloured paper (*Haderlumpen*) were manufactured as packing paper which could be employed for wrapping fine wire articles in 'needle packets', as cardboard and also as material for strengthening playing cards. Playing cards are first heard of in Nuremberg in 1370 simultaneously with the earliest notices in Florence and Basle, all in connection with the prohibition of the new evil epidemic of card play. Printing in movable type and the deep impression required for copperplate engraving

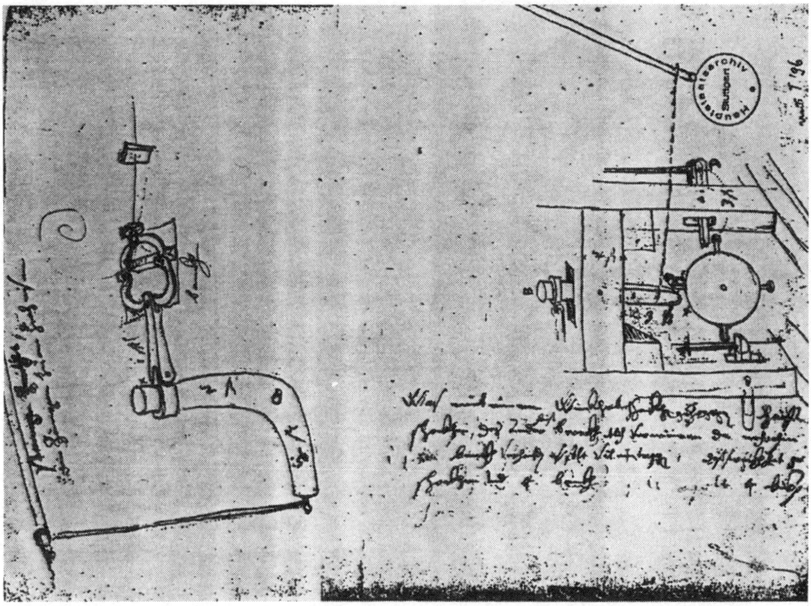

Figure 11 The fully automatic wire-drawing machine. Details from a drawing of 1596. *Source:* Hauptstaatsarchiv Stuttgart.

required absolutely even-quality paper and correspondingly fine mesh sieves, with wire as fine as hair and completely uniform in diameter (Figure 12). Increased production of paper and wire was mutually dependent. Just as mutually reinforcing waves of innovations were characteristic of the great industrial revolution of modern times, so they had forerunners at the close of the Middle Ages.[30]

(vii) I shall omit here the most popular Nuremberg invention apart from the famous funnel, namely, the pocket watch shaped in the form of an egg made by Peter Henlein. A certain Peter Henle (*c.* 1485–1542) achieved great recognition for his self-operating watch. That he was its inventor, however, cannot be proven, and is improbable. On the other hand, the folding combined compass-sundial was a mass-produced speciality of Nuremberg which had been in great demand for generations and was of the type which had earlier been produced by the *orloymeister bei dem Stromeyr*.[31]

Until now scholars have regarded Nuremberg as remarkably backward in the development of printing. Half a generation passed between Gutenburg's production of the Cyprus indulgence of 1453–4 and the 42-line Bible, and the establishment of the first printing press in Nuremberg in 1469. The presses of Strasbourg, Bamberg, Eltville, Cologne, Subiaco, Rome, Augsburg, Basle and Venice of 1460–69 were all earlier than

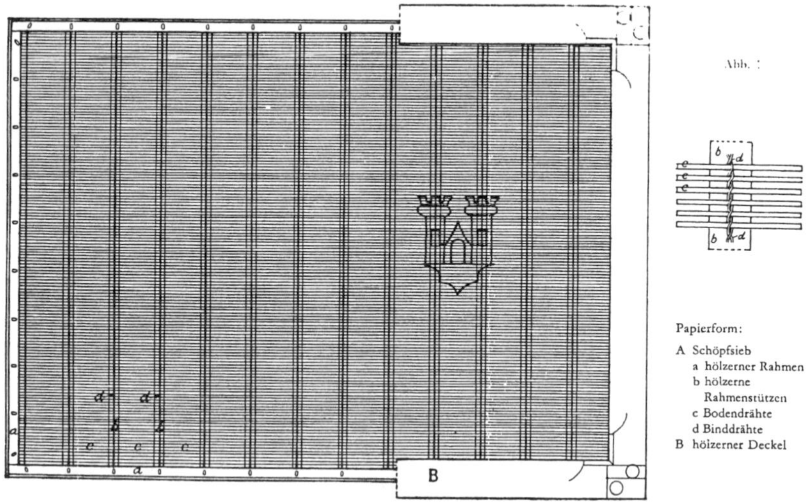

Figure 12 Form and construction of a wire sieve for paper-making of the fifteenth century. *Source*: G. Piccard, 'Die Wasserzeichenforschunge als historische Hilfswissenschaft', *Archivalische Zeitschrift*, 1956, 52:67, fig. 1.

Nuremberg's. Nevertheless, the first Nuremberg printer, Johann Sensenschmid in 1469 'from Eger' (today Cheb, in western Bohemia) came from a Nuremberg family in the wire business and not from the Bohemian city, which was only his last place of residence. His partner, Heinrich Keffer, had been an assistant of Gutenberg in 1455. In 1470, a year later than Sensenschmid, Anton Koberger opened his printing works, which soon became the most important printing and publishing house in Europe, and remained so for three decades. Shortly afterwards the genius Regiomontanus opened his own printing house to publish his own and others' most recent research, and – for the first time – printed instructions for the apparatus which he had built.[32]

The slight delay in the establishment of book printing with movable type in Nuremberg was compensated for by the city's pioneer role in the history of the invention of printing itself. Of all the playing cards which after about 1370 flooded a world which soon became addicted to card games, no single playing card survives from before 1436, let alone a complete pack. Three out of the five earliest cards to survive are printed on paper datable from the watermarks of the *Hadermühle* (the rag pulping mill) which belonged to the Stromeirs (Figure 13). They are a death of Mary and a St Christopher, both printed on paper made in 1422, and a third showing two angles standing before a monstrance (Figure 14). On the last we have also the first printed Western text, namely the words '*Ecce panis angelorum*' on paper which by its watermark can be shown to have been made in 1427.[33]

From 1433 to his death in 1459, Conrad Forster, a Dominican monk

Figure 13 Stromeir paper-mill watermarks of the period 1392–1453.

from Nuremberg, sacristan of the Dominican church, printed 85 long texts (printed with stamps) on the leather bindings of his books in his abbey and on those of the nunnery of St Catherine which belonged to the order. These bindings have survived until today, and in addition 70 other volumes with decorative bindings by Forster are also extant (Figure 15). Until recently they had been scarcely noticed because the method of printing with stamps was blind printing (by impression only) without ink.

Set beside the superlative achievement of Gutenberg, with its wonderful black lettering, it has been seen as a failed innovation of secondary importance, a blind alley which led nowhere. The experts failed to notice that two of the Gutenberg bibles, still in their original leather binding from Mainz, also carried such indented printed letters, alternating with little stars printed in *Hochdruck* (indented printing). Since Gutenberg is known to have been very businesslike in the sale of his books, we can presume that he was responsible for the form and appearance of the binding. There is a plausible explanation of how he learned about the Nuremberg method of printing with stamps. There were numerous contacts between the nuns of the Nuremberg house and nuns of a house in Alsace, where women from the families of Gutenberg's associates were members.[34] The singular nature of the forms of Forster's letters and those of the *Stempeldrucker* in Mainz (that is *gespornten Lettern* – literally 'bespurred letters', meaning 'letters with tails', i.e. e, f, g, r, t and x) permits one to say that they are characteristic of the Nuremberg goldsmith Hensel Sigerstorfer, who between 1430 and 1435 used such forms for the first time in the history of experimental printing . Proof of this is that he can be identified as the first experimenter, since he engraved a relatively long text with his name on the bronze wings of a small altar and advertised his skill with a small early copperplate

Figure 14 Two angels holding a monstrance. Woodcut on paper from the Stromeir paper-mill bearing a watermark of 1427 (the ox-head of Figure 13).

engraving with three alphabets (Figure 16), each of which is identical to those of Forster and where 'letters with tails' also appear.[35]

CONCLUSION

Each of these inventions made its contribution to the economy of Central Europe. North of the Alps each of the innovations described here was first employed in Nuremberg and contributed to the well-being of the city, its citizens and its crafts. The fact that a large number of the inventions were basic and pioneer inventions, having a cascade effect which stimulated waves of further innovation, was important. This made Nuremberg the epicentre of what one might even describe as an 'industrial revolution of the later Middle Ages'.[36]

One might well ask what made such developments possible, since Nuremberg, situated on an unimportant river and without any worthwhile deposits of ores, would not seem to have been predestined

Figure 15 A book cover with punched lettering of 1440 by Conrad Forster.

Figure 16 Hensel Sigerstorfer's alphabets with 'spurred' letters of 1430–5.

to play such a role. From the standpoint of geographical position on the main transport axis of Europe, Bamberg or Regensburg would seem to have enjoyed greater chances. The fact is that Nuremberg's advantages were man-made and the result of the system of bilateral and multilateral tariff-free agreements which she negotiated with other cities. The patrician oligarchy which dominated the city from the Staufer period to the end of the old Empire had never permitted guilds, whose restrictive practices might have hindered progress. The merchants of Nuremberg enjoyed credit and their products were trusted because the city was known to be law-abiding and to respect the rule of law. It was the first city to gather its legal practices into a great codex, and even printed this in 1484, thus making it available for inspection by all. In the above-mentioned privilege which the Nurembergers received from Heinrich VII on 11 June 1313, it was laid down that a judge should make his decisions without regard to whether the parties were rich or poor, 'tam pauperibus quam divitibus'. The ruling oligarchy remained true to this principle and was prepared even to sacrifice its own private interests to those of the city and its fellow citizens, as is evidenced by the closure of the smelting and liquation furnaces, the *Saigerhütten*, mentioned earlier. They regarded themselves as equally subject to the laws they enacted and administered. Nowadays this seems absurd.[37]

Notes and References

This article is a translation by Natalie Fryde of 'Nürnberg als Epizentrum von Erfindungen und Innovationen an der Wende vom Mittelalter zur Neuzeit', in Karl Albrecht Schachtschneider (ed.), *Wirtschaft, Gesellschaft und Staat im Umbruch: Festschrift der Wirtschafts- und Sozialwissenschaftlichen Fakultät der Friedrich-Alexander-Universität Erlangen-Nürnberg 75 Jahre nach Errichtung der Handelshochschule Nürnberg* (Berlin 1996), 668–87. The second part of the translation, 'Technological inventions and innovations', has been revised by Dr Graham Hollister-Short.)

1. Gerhard Hirschmann (ed.), 'Nürnbergs Handelsprivilegien, Zollfreiheiten und Zollverträge bis 1399', *Beiträge zur Wirtschaftsgeschichte Nürnbergs* (BWGN), I/1967, 1-48, here nos. 38/1332, 10/1227, 14/1304, 19/1313; Hektor Ammann, 'Die wirtschaftliche Stellung der Reichsstadt Nürnberg im Spätmittelalter', *Nürnberger Forschungen*, 1970, 13: 20–5; W. von Stromer, 'Handel und Gewerbe der Frühzeit Nürnbergs', in Gerhard Pfeiffer (ed.), *Geschichte einer europäischen Stadt* (Munich, 1971), 46–54, and 'Nürnbergs große Zollfreiheiten, ihre Symbole und ihre Monumente', *Mitteilungen des Vereins für die Geschichte der Stadt Nürnberg* (MVGN), 1993, 80: 117–35 (with map).
2. Hirschmann, *op. cit.*(1), 'Handelsprivilegien ...', nos. 5, 6/1264, and 19/1313.
3. Hirschmann, *op. cit* (1), 18/1311 (Brabant), 36/1326 (Bohemia), 45,46/1336, 47/1337, 74,75/1357, 99,100/1364, 127–130/1383 (Hungary), 90–95/1362 (Flanders), 101/1365 (Poland and Galicia), 160/1398 (Genoa); W. von Stromer, *Oberdeutsche Hochfinanz 1350–1450*, 3 vols, Beihefte 55–57 zur VSWG, Wiesbaden 1970, here ch. 2, 'Acquisition of the great Flemish Privilege of 1362'; ch. 3, 'The German Merchant in Milan' (p. 51: the Genoese provisions of 1346); ch. 4, 'The Penetration of German Entrepreneurs of the Carpathian Region'.
4. W. von Stromer, 'Die Metropole im Aufstand gegen Karl IV: Nürnberg zwischen Wittelsbach und Luxemburg Juni 1348–September 1349', MVGN, 1978, 65: 55–92 (for Nuremberg and Venice 1340–48, 1358).
5. W. von Stromer, 'Landmacht gegen Seemacht: Kaiser Sigismunds Kontinentalsperre gegen Venedig 1412–1433', *Zeitschrift für Historische Forschung* (ZHF), 1995, 22: 145–89.
6. Henri Simonsfeld, *Der Fondaco dei Tedeschi in Venedig und die deutsch-venetianischen Handelsbeziehungen (1225–1653)*, 2 vols (Stuttgart, 1887), Vol. 1, nos. 240, 307, 351, 261–4, 821 (= 363a), pp. 109, 158, 184, 192–4, 452; Vol. II, pp. 203–12, List of the consuls 1492–1753.
7. Hermann Aubin, 'Formen und Verbreitung des Verlagswesen in der Altnürnberger Wirtschaft', BWGN, 1967, Vol. II: 620–88; W. von Stromeir, 'Der Verlag als strategisches System einer an gutem Geld armen Wirtschaft', VSWG, 1991, 78: 153–71, and 'Une clé du succès des maisons de commerce d'Allemagne du Sud: le grand commerce associé au *Verlagssystem*', *Révue Historique*, 1991, 285: 29–49. On the *Verlagssystem* in general, see Rudolf Holbach, *Frühformen von Verlag und Großbetrieb in der gewerblichen Produktion* (Stuttgart, 1993).
8. W. von Stromer, 'Die Große Oberpfälzer Hammereinigung vom 7. Januar 1387: Kartell und Konzerne, Krisen und Innovationen in der mitteleuropäischen Eisengewinnung', *Technikgeschichte*, 1989, 56: 279–304, and 'Cartels, consortiums et innovations dans l'exploitation du fer d'Europe Central à la fin du Moyen Age', in Paul Benoit and Pierre

Fluck (eds), *Techniques Minières, 113th Congress Nat. Soc. Savantes Strasbourg 1988* (Paris, 1992), 167–86.

9. Ulman Stromeir, 'Püchel von meim geslecht und von abenteur', in Karl Hegal (ed.), *Die Chroniken der deutschen Städte*, Vol. I (Leipzig, 1862), 1–312, here pp. 77–83 on the paper-mill; W. von Stromer, 'Ulman Stromeir, Leben und Leistung: 81 Dokumente zur Geschichte der Stromeir'schen Papiermühle 1390–1453', in Lotte Kurras (ed.), *Kommentarband zur Faksimileausgabe des Püchel*' (Bonn, 1990), 89–170, pp. 149, 152 *et seq*. Regesten 7b, 15 and 22 for the *Verlags* agreement with J. Tirmann.

10. W. von Stromer, *Die Gründung der Baumwollindustrie in Mitteleuropa, Spätmittelalterliche Wirtschaftspolitik*. Monographien zur Geschichte des Mittelalters, Vol. 17 (Stuttgart, 1978), pp. 30 *et seq*., 100–26, and 'Gewerbereviere und Protoindustrien in Spätmittelalter und Frühneuzeit', in Hans Pohl (ed.), *Gewerbe- und Industrielandschaften vom Spätmittelalter bis ins 20. Jahrhundert*, Beiheft 68 of VSWG (Stuttgart, 1986), 39–111, 63–7, 70–2.

11. W. von Stromer, 'Organisation und Struktur deutscher Unternehmen in der Zeit bis zum Dreißigjährigen Kreig', *Tradition*, 1968 13: 29–37, and 'Zur Struktur der Handelsgesellschaften in Oberdeutschland', in Frederick Lane (ed.), *Troisième Conférence Internationale d'Histoire Economique (Munich, 1965)*, Vol. V (Paris and The Hague, 1974), 153–6, 259–62.

12. W. von Stromer, 'Das Schriftwesen der Nürnberger Wirtschaft vom 14. bis zum 16. Jahrhundert: Zur Geschichte Oberdeutscher Handelsbücher', BWGN 1967, II (as for note 1), 751–99; 'Tuchhandel im Spiegel oberdeutscher Handelsbücher' in Marco Spallanzani (ed.) *Produzione, Commercio e Consumo dei Panni di Lana*, Istituto Francesco Datini Prato. Atti II/2, Florence, 1976, 325–40, and 'Binationale deutsch–italienische Handelsgesellschaften im Mittelalter', in Josef Riedman (ed.), 'Kommunikation und Mobilität im Mittelalter', *Sigmaringen*, 1995, 135–59.

13. U. Stromeir, *Püchel, op. cit.* (9), 77–83, 99–106; W. von Stromer, 'Schriftwesen . . .', *op. cit.* (12), 781–5; 'Hochfinanz' *op. cit.* (Note 3), 74 *et seq*., II, 480 *et seq*.; and 'Vom Stempeldruck zum Hochdruck – Forster und Gutenberg', in Holger Nickel and Lothar Gillner (eds), *Johannes Gutenberg: Regionale Aspekte des frühen Buchdrucks* (Berlin, 1993), 47–92, p. 72; Erich Meuthen, 'Ein frühes Quellenzeugnis über den ältesten Buchdruck', *Gutenberg-Jahrbuch 1982*, 108–18.

14. W. von Stromer and Michael Toch, 'Zur Buchführung der Juden im Spätmittelalter', in Jürgen Schneider (ed.), *Wirtschaftskräfte und Wirtschaftswege: Festschrift für Hermann Kellenbenz*, Vol. I, (Stuttgart, 1978), 387–412, and 'Hartgeld, Kredit und Giralgeld: Zu einer monetären Konjunkturtheorie des Spätmittelalters und der Wende zur Neuzeit', in Vera Barbagli Bagnoli (ed.), *La Moneta nell'economia europea secoli XIII–XVIII*, Istituto Francesco Datini Atti II/7 (Prato, 1981), 105–25, 144–61.

15. W. von Stromer, 'Funktion und Rechsnatur der Wechselstuben als Banken in Oberdeutschland, den Rheinlanden und den mitteleuropäischen Montanzentren im Spätmittelalter', *Bankhistorisches Archiv, Zeitschrift zur Bankgeschichte*, 1979, 57: 3–34; 'Banken und Geldmärkte: Die Funktion und Rechtsnatur der Wechselstuben als Banken in internationalen Vergleich', in Anna Vannini-Marx (ed.), *Credito, Banche e Investimenti, sec. XIII–XIV*, Istituto Francesco Datini Prato, II/4 (Florence, 1985), 229–54; and 'Die oberdeutschen Geld- und Wechselmärkte: Ihre Entwicklung vom Späatmittelalter bis zum Dreißigjährigen Krieg', *Scripta Mercaturae*, Vol. 10, 1976, 23–49; Raymond de Roover, *Money, Banking and Credit in Medieval Bruges* (Cambridge, MA, 1948), Part III, *The Money Changers*, 171–344.

16. Heinrich Brunner, 'Das erste deutsche Rechenbuch', MVGN, 1937, 35: 1–16; Wolfgang Schweiker (pseudonym), *Zwifach Buchhalten sampt dem Giornal* (Nuremberg (J. Petreius), 1549); Lorenz Meder, *Handelbuch* (Nuremberg, 1588); Hermann Kellenbenz, *Deutsche Handelsakten des Mittelalters und der Neuzeit*, Vol. 15 (Wiesbaden, 1974); Georg Nicolaus Schurtz, *Nutzbare Richtschnur der löblichen Kaufmannschaft* (Nuremberg, 1662, reprinted 1695); W. von Stromer, 'Oberdeutsche Geld . . .', *op. cit.* (15), with numerous examples in facsimile.

17. Max Liedtke, 'Johann Neudörfer d. Ä., Schreibmeister und Historiograph', in Christof Imhoff (ed.), *Berühmte Nürnberger aus neun Jahrhunderten* (Nuremberg, 1989), 119 *et seq*. Neudörfer's *Nachrichten* was printed for the first time in 1828. Johann Gabriel Doppelmayr's *Historische Nachricht von den Nürnberger Mathematicis und Künstlern* (Nuremberg, 1730) was unique for its time in dealing with such subject matter.

18. Regiomontanus, *Vsum Ephemeridi* (Nuremberg, 1474) (Year Books of star positions for

the years 1475–1531). Reprinted by Erhard Ratdolt (Venice, 1481) a copy of which survives with annotations by Columbus. On this see Ernst Zinner, *Leben und Wirken des Johannes Müller von Königsberg, genannt Regiomontanus*, 2nd ed. (Osnabrück, 1968), 189 and 300, table 34, illustration 59; Angelika Wingen-Trennhaus, 'Regiomontanus als Frühdrucker in Nürnberg', MVGN, 1991, 78: 17–87; W. von Stromer, 'Hec Opera fient in oppido Nuremberga Germanie ductu Joannis de Monteregio', in Günther Hamann (ed.), *Regiomontanus-Studien*, Akad. d. Wiss. Phil. Hist. Kl. SB Vol. 364 (Vienna, 1980), 267–89, and 'Meister Konrad Scherp, Regiomontanus Experte ...', MVGN, 1992, 79: 123–32.

19. W. von Stromer, 'Hec Opera fient ...' *op. cit.* (18), 267–79; J. Hartmann, 'Die astronomischen Instrumente des Kardinals Nicolaus Cusanus', *Abhandl. d. kgl. Gesellsch. d. Wiss. zu Göttingen, math.-phys. Kl.*, NS, 1919, 10(6): 1–50.

20. Since Copernicus's main work appeared in Nuremberg in 1543, the German festivities celebrating the 500th anniversary of his birth took place there. The German government of the day threatened to cut all subsidies for the festivities if the question of his nationality was as much as mentioned. Memorials have been erected to the 'greatest son of Poland' throughout the world. Nevertheless, he was born in Thorn (now Toruń) of German parentage in 1473. After 1466 the town belonged to the royal part of Prussia. The Polish scholar Marian Biskup published his work, *Copernicus, Prussia's Greatest Son*, in 1973, and this was passed by the Polish censors. I had been co-opted on to the Copernicus Committee at the request of Polish colleagues but was removed by German representatives, who also prevented me from using the Copernicus Research Centre of the Deutsches Museum in Munich.

21. Herbert Krüger, 'Erhard Etzlaub, Kartograph, Uhren- und Kompaßmacher ca. 1460–1532', in Imhoff (ed.), *op. cit.* (17), 76 *et seq.* Presumably printed for the 'Holy Year 1500', his map of the route to Rome located 558 places in the correct order. His map of Germany of 1501 located as many as 830 places!

22. Albrecht Dürer, *Unterweysung der messung mit dem Zirkel und Richtscheyt* ... (1525), *Etliche Unterricht zur Befestigung der Stett, Schloß und Flecken* (1527) and *Vier Bücher von menschlicher Proportion* (1528); Paul Pfinzing, *Atlas* (1594) (Facsimile, Peter Fleischman (ed.), Staatsarchiv Nürnberg, 1994), and *Methodicus Geometrica* (Nuremberg, 1598); Ernst Gagel and Fritz Schnelbögl, *Pfinzing, der Kartograph der Reichstadt Nürnberg (1554–1599)* (Hersbruck, 1957).

23. W. von Stromer, 'Ein Lehrwerk der Urbanistik aus der Spätrenaissance: Die Baumeisterbücher des Wolf-Jacob Stromer (1561–1614), Ratsbaumeister zu Nürnberg', *Willibald-Pirckheimer-Gesellschaft, Year Book* Vol. I (Nuremberg, 1984), and in August Buck and Bodo Guthmüller (eds), *La Città Italiana del Rinascimento fra Utopia e Realtà* (Centro Tedesco di Studi Veneziani, Quaderni 27, Venice, 1984), 71–115. See also J. Needham, *Science and Civilisation in China* (Cambridge, 1971), Vol. 4, Part 3, 61 *et seq.* but especially 175 and Fig. 835. The segmental arch bridge appears first in the West in the late thirteenth century but was applied largely in the fourteenth, e.g. the Ponte Vecchio, Florence, of 1345. The maximum span was that at Trezzo, 1375, a span of 74 m. The best example of the type in China is the An-Chi Bridge in southern Hopei of +610, with a span of 37.8 m.

24. W. von Stromer, 'Wassersnot und Wasserkünste im Bergbau des Mittelalters und der frühen Neuzeit', in Werner Kroker and Ekkehard Westermann (eds), *Montanwirtschaft Mitteleuropas vom 12. bis zum 17. Jahrhundert, Der Anschnitt*, Supplement 2 (Bochum, 1984), 50–73.

25. Lore Sporhan and W. von Stromer, 'Die Nadelholz-Saat in den Nürnberger Reichswäldern zwischen 1469 und 1600', ZAGAS (*Zeitschrift für Agrargeschichte und Agrarsoziologie*), 1969, 17: 79–109; Georg Sperber, 'Der Reichswald bei Nürnberg: Aus der Geschichte des ältesten Kunstforstes', *Mitteilungen der Staatsforstverwaltung Bayerns*, Vol. 37 (Munich, 1968); W. von Stromer, 'Die Erfindung der Nadelwald-Saat durch Peter Stromeir von Nürnberg 1368', *Schöpferische Leistung, zur 13. Verleihung der Dieselmedaille* (1968), 58–61.

26. W. von Stromer, 'Hammereinigung', *op. cit.* (8), 294–7; 'Ulman Stromeir', *op. cit.* (9), Regesten 18, 62, 81, p. 152, 164 *et seq.*, 169, and sketch map (of 1601!), 167; and 'Bemessung der Energie Wasserkraft', in Rainer S. Elkar *et al.* (eds), *Vom rechtem Maß der Dinge: Festschrift für Harald Witthöft zum 65. Geburtstag* (Siegen, 1995), 127–44.

27. Lothar Suhling, *Der Seigerhüttenprozess: Die Technologie des Kupferseigerns nach dem frühen metallurgischen Schrifttum* (Stuttgart, 1976); Ekkehard Westermann, *Das Eislebener Garkupfer und seine Bedeutung für den europäischen Kupfermark 1460–1560* (Cologne and Vienna, 1991); W. von Stromer, 'Die Saigerhütten-Industrie des Spätmittelalters: Entwicklung der Kupfer–Silber

Scheidekunst zur "ars conflatoria separantia argentum a cupro cum plumbo"', *Technikgeschichte*, 1995, 62(3): 187–219.

28. W. von Stromer, 'Große Innovationen der Papierfabrikation in Spätmittelalter und Frühneuzeit', *Technikgeschichte*, 1993, 60(1): 1–6 for scissors; Bartolomeo Paxi, *Tariffa de Pexi e mesure* (Venice, 1503) (six eds to 1560) for the 'Merze de Fontego'.

29. W. von Stromer, 'Innovation und Wachstum im Spätmittelalter: Die Erfindung der Drahtmühle als Stimulator', *Technikgeschichte*, 1977, 44: 89–120, and 'Apparate und Maschinen von Metallgewerben in Mittelalter und Frühneuzeit', in Harry Kühnel (ed.), *Handwerk und Sachkultur im Spätmittelalter*, Österr. Ak. d. Wiss., phil.-hist. Kl., SB Vol. 513 (Vienna, 1988), 127–49, pp. 138–41 and illus. 18a–20 for the wire mills.

30. W. von Stromer, 'Innovation der Papierfabrikation', *op. cit.* (28).

31. W. von Stromer, 'Hec Opera', *op. cit.* (18), 279; Penelope Gouck, *The Ivory Sundials of Nuremburg 1500–1700* (Cambridge, 1988); for Peter Henlein, Watchmaker, see Johannes Willers in Imhoff (ed.) *Berühmte Nürnberger*, *op. cit.* (17), 105 *et seq.*

32. Eduard Isphording, 'Johann Sensenschmid, Buckdrucker', in Imhoff (ed.), *Berühmte Nürnberger*, *op. cit.* (17) 66 *et seq.*; Hermann Glaser, 'Anton Koberger, Drucker und Großverleger, *ibid.*, 55–7; Oskar Hase, *Die Koberger, Darstellung des Buchhändler-Geschäftsbetriebs* (Leipzig, 1885); Regiomontanus, *Vsum Ephemeridi*, *op. cit.* (18).

33. Illustrations of the playing cards of 1436, of the woodcut of the Death of Mary and the St Christopher of 1422, in Jürgen Franzke and W. von Stromer (eds), *Zauberstoff Papier: Sechs Jahrhunderte Papier in Deutschland*, 33–4; for the woodcut of the 'Engel mit Monstranz und Spruchbänder von 1427', see W. von Stromer, 'Die erste Papiermühle in Mitteleuropa . . .', in Simonetta Cavaciocchi (ed.), *Productione e Commercio della Carta e del Libro sec. XIII–XVIII* (Istituto F. Datini in Prato, II/23, Florence, 1992), 297–310, p. 306.

34. W. von Stromer, 'Vom Stempeldruck' *op. cit.* (13), and 'Fränkische Buchkultur zur Gutenbergzeit, *Jahrb. für Fränk. Landesforschung*, Vol. 52, 1992 (Festschrift für Alfred Wendehorst), 349–66.

35. W. von Stromer, 'Gespornte Lettern: Leitfossilien des Stempeldrucks, c. 1370–1490', *Gutenberg Jahrbuch 1996*, 23–64. It is remarkable that the six letters with tails – e, f, g, r, t, x – are constructed by mounting the tails on various parts of the letters c, s, b, i, l, r: apparently an attempt at rationalization.

36. W. von Stromer, 'Pionier-Innovationen und Innovationsschübe und ihr Einfluß auf Wirtschafts- und Lebensbereiche', in Harry Kühnel (ed.), *Mittelalter und Frühneuzeit*, Österr. Ak. d. Wiss. phil.-hist. Kl., SB Vol. 470 (Vienna, 1986), 121–35, and 'Eine "Industrielle Revolution" des Spätmittelalters?', in Ulrich Troitzsch and Gabrielle Wohlauf (eds), *Technik-Geschichte* (Frankfurt, 1980) (Suhrkamp TB Wiss. 319), 105–38.

37. Lazarus Karl von Wölckern, *Historia Norimbergensis Diplomatica* (Nuremberg, 1738), 227, No. 71, Kaiser Henrici VII, Freyheiten und Verordnungen . . . im Jahre 1313.

Friction According to Jacob Leupold

WILFRED G. LOCKETT

In his seminal paper on 'mirror-image twins'[1] Edwin Layton advances the notion that engineers have developed their own special way of dealing with technological problems – an approach which is analogous to but distinct from conventional 'science'.

But is Layton's thesis limited by its specific reference to nineteenth-century America? It could be argued that the 'science' of engineering was already quite well developed by the time North American industrial expansion got under way in the first decades of the nineteenth century. Indeed, few would deny that the gradual adoption of the scientific method by engineers was initiated in Europe a hundred years previously, technological 'rationalism' and 'empiricism' appearing at the same time, depending on the background (and country) of the participants. The pioneers in this movement included not only 'scientists' with an interest in technology, but also some non-academic practitioners, who recognized the need to apply the fundamentals of mechanics rather than traditional rules of thumb. Among the latter was Jacob Leupold, who flourished in Leipzig in the first decades of the eighteenth century.

According to what is almost certainly the earliest account of his life,[2] Leupold was born on 25 July 1674 in Planitz, near Zwickau. His father, George Leupold, was a man 'well-experienced in the trades of cabinet-maker, turner, sculptor and watchmaker'. After he had attended the Zwickau school Jacob's subsequent education was evidently informal. It is clear that from periods under reputed professors at the universities of Jena, Wittenberg and Leipzig, he acquired a good grounding in mathematics – sufficient to enable him to support himself by giving private lessons in this subject. It was for these lessons that he started making mathematical instruments, and this in due course became his primary business. The 'instruments' (the word must be interpreted broadly) produced included *inter alia* weighing-scales, air-pumps, water gauges, recording rain-gauges, barometers, and surveying and draughting instruments. He was associated with the University of Leipzig as 'Mechanic', a position which invites comparison with that of James Watt at Glasgow.[3] He became a member of the Berlin Academy of Sciences on 12 April 1715. His reputation as inventor, instrument and model-maker brought him to the attention of the courts of Saxony and Prussia, where he was appointed, respectively,

Commissioner of Mines and Counsellor of Commerce (*Commercien-Rath*). He died suddenly in Leipzig on 12 January 1727.

In an article on the status of research on Leupold,[4] Ulrich Troitzsch draws our attention to his most notable achievement, the *Theatrum Machinarum*, a ten-volume encyclopaedia of technology which was published in Leipzig during the years 1724–39 (Volumes VIII, IX and X posthumously).[5] In his opening preface to Volume I, the *Theatrum Generale*, Leupold makes it quite clear that the prime purpose of his work was the exposition, in the vernacular, of the basic principles of mechanics

> not to the learned and experienced mathematicians [*Mathematici*] who are already, or should be, better acquainted with them ... [and most of whom] have studied mechanics more as a subject of curiosity and a hobby, than with any view of service to the public. The people we had in mind were rather the mechanic, handicraftsman [*Künstler, Handwercker*] and the like, who, without education or knowledge of foreign languages have no access to many sources of information.[6]

Leupold is here drawing a clear distinction between, on the one hand, the scholarly mathematician who knew Latin and often took a lively interest in technology mainly from the theoretical point of view and, on the other, the skilled artisan whose education was severely limited. His own schooling having evidently left him with a reasonable knowledge of mathematics (as well as Latin), and with his family background and subsequent practical experience, Leupold described himself as a *mechanicus* – one who, while fully conversant with the principles of mechanics, nevertheless understood their application in the real world. We may take *mechanicus* as the equivalent of 'engineer' in the modern sense of the word.

To demonstrate his theme Leupold wanted his work to be considered as a *Schau-Platz*; that is, a 'visible display'. And indeed, as a highly illustrated work it shares many characteristics with the other similarly titled volumes that preceded it. But Alex Keller takes a more narrow view of the 'theatres of machines' genre.[7] For him the archetypes were the single-volume publications which appeared during the latter years of the Renaissance, as represented by those of Besson (1594), Ramelli (1588), Zonca (1607) and Veranzio (1615).

A common feature of these early *Theatra* was that the illustrations, typically of whole machines (as opposed to mechanical components), themselves determined the format of the book. In most but not all cases the picture and its description were juxtaposed. The illustrations were often quite skilfully engraved, as much with a view to artistic merit as to technical accuracy.

The range of mechanical devices shown in these *picture-books* was enormous. The impression given is that any new invention was worth showing, whether or not it had any practical use, but particularly if it displayed 'cleverness'. For example, as well as canal locks having mitre gates (Zonca), horizontal windmills with curved vanes and self-anchoring arch bridges (Veranzio), we find capstans with impracticable combinations of screws and pulleys (Ramelli) and perpetual-motion devices (Zonca).

Leupold himself, in pursuing his objective of showing the 'state of the art' as it stood in the early eighteenth century, clearly felt obliged to incorporate many of the machines illustrated by his predecessors. Thus, with their numerous illustrations, Leupold's volumes may also be regarded as *picture-books*. Furthermore, his explanations of mechanical principles are often presentations of the work of the original authors, with frequent quotations and due acknowledgement of his sources.[8] The format, however, is determined by the subject matter and not the *Tabulae*, and, particularly in the *Theatrum Generale*, the tendency is to illustrate mechanical components rather than illustrations of whole machines, all with his didactic objective in view.[9]

In the course of his paper,[10] Troitzsch draws attention to Leupold's judgement of the work of his predecessors, including his scornful rejection of the *perpetuum mobile* and, in particular, his criticism of over-elaborate machines (such as certain devices illustrated by Ramelli[11]), on account of the excess friction involved. Leupold's treatment of friction merits our attention because it is an early example of the use of 'basic principles' (such as were established at the time) to explain a common problem in the operation of machines.

The chapter on friction is located in the *Theatrum Generale*, between the two main parts of that volume. In Chapters I to XV Leupold discusses what he terms 'the internal forces in mechanics'. Here he describes the classical elements (various levers, the wedge, etc.) and, using the methods of taking moments and the resolution of forces, shows how pulley systems, gear trains, cams and screws can be arranged to increase the force needed to move a load. Concluding this section in Chapter XV, Leupold is at pains to point out that allowance has to be made for the effect of friction, which had been ignored in the preceding analyses. After his discussion of friction in Chapter XVI (which is the main concern of this paper) Leupold continues with the second part of the volume, on the 'external forces in mechanics', in which he considers a wide range of prime movers, with power sources classified as 'animal', 'wind', 'fire', 'water' and 'springs'.

Leupold's definition of friction opens Chapter XVI thus:

> Our understanding of the resistance to movement, known in Latin as *Frictio*, is as follows: when the various parts of a machine are in relative motion the adjacent surfaces such as axles or gudgeons drag and grind against each other, due to their unevenness in the form of small cavities, into which the surfaces fall and from which they must be raised. This requires a force which might otherwise have been usefully employed in moving the load.
>
> The heavier the weight on the surfaces and bearings the greater will be the mutual pressure, so that correspondingly more force will be required to achieve movement. From this it follows that:

> The resistance or friction will increase as the roughness and unevenness of the surfaces and pressure between them increases, and, conversely, the friction will decrease as the pressure decreases and the surfaces are smoother.

Leupold then asks rhetorically whether or not the frictional force is affected by the area of contact of the surfaces, and refers to the experiments conducted by Amontons,[12] which demonstrated (at least to Amontons's satisfaction) that it does *not* depend on the area but does indeed depend on the imposed weight. He then cites a paper by Sturm[13] which challenges this assertion, saying that if it were true a mill axle would move just as easily in a 6-inch journal as in a 3-inch journal (which is obviously not the case). But Leupold sees the flaw in Sturm's argument, and shows that the difference in the two cases is due to the fact that the frictional resistance of the journal manifests itself as a *torque*. In other words, the frictional force is the same, but, as it acts tangentially to the contact surfaces of the journal, it has a moment about the axis which is dependent on the radius of the journal itself (thus the bigger the journal the greater the resistance to turning).

Leupold goes on to describe Amontons's experiments with boards (Figure 1), in which the force required to move the boards was $\frac{1}{3}$ of the load (ranging from 30 lb, 60 lb to 300 lb), whether the boards were placed flat, on edge, or on a reduced area (as A, E and K respectively in Figure 1). Leupold himself repeated the experiments and obtained results which varied from Amontons's 'by barely a half or a whole pound'. The proportion $\frac{1}{3}$ (which Amontons maintained was applicable in all circumstances) is of course our familiar *coefficient of friction* (commonly in modern notation μ).

There follows a description of Leupold's experiments with 'the roller or gudgeon' (pp. 99–100):

> [The conclusion reached by Amontons] also applies, therefore, in the case of a gudgeon or bearing, whatever its diameter. It is denied by Sturm as already discussed above, but it appears from the example of the mill wheel that Sturm had no real idea of the matter, even though he claimed to have performed such an experiment. Amontons's concept must be understood as I have shown by the apparatus in [Figure 2], where in the frame AB there are two trunnion bearings C and D, in which lies a cylinder or spindle E having a uniform diameter of about 3 inches. Suspended from flexible cords wound round the shaft are two weights of 30, 60 or more pounds. Now for the one weight to turn the spindle it is necessary to increase it by $\frac{1}{3}$. Since, having placed on each side 30 lb, I must attach a further $10\frac{1}{2}$ lb, so, with 60 lb on each side, I must add 21 lb, and subsequently (perhaps because the spindle was smoother) only 20 lb was necessary. The same as obtained with the 3-inch spindle was also found with one of 6 inches in diameter ...

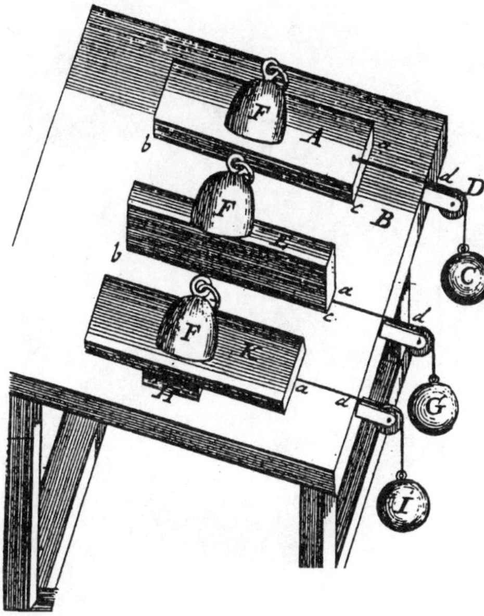

Figure 1 Amontons's apparatus.

> I have found from a variety of [similar] trials that ⅓ the weight (plus or minus about 1 lb) is required to move the load. Thus the friction is the same, whatever the area of the bearing or the diameter of the spindle, increasing only in proportion with the load. But we cannot derive from this a universal law that in every machine ⅓ of the imposed load is required to overcome the friction.
>
> Regarding friction one must take into account:
> 1. the material of the surface;
> 2. the roughness or unevenness of the surface;
> 3. the shape of the uneven parts and their depth and height;
> 4. the line of action of the force relative to the direction of motion.

In this quotation Leupold is demonstrating, with the aid of his experiments, the application of Amontons's results in the operation of machine components found in practice (a significant proportion of the devices he is concerned with involve spindles turning in bearings). After noting above that friction varies in accordance with the materials involved, their hardness and degree of finish, Leupold goes on to discuss in greater detail the effect of surface texture (§218, §219):

> There is similarly a great difference in the roughness or unevenness of surfaces, rough boards from the sawmill requiring almost half the weight to overcome the friction. When the peaks and hollows in the

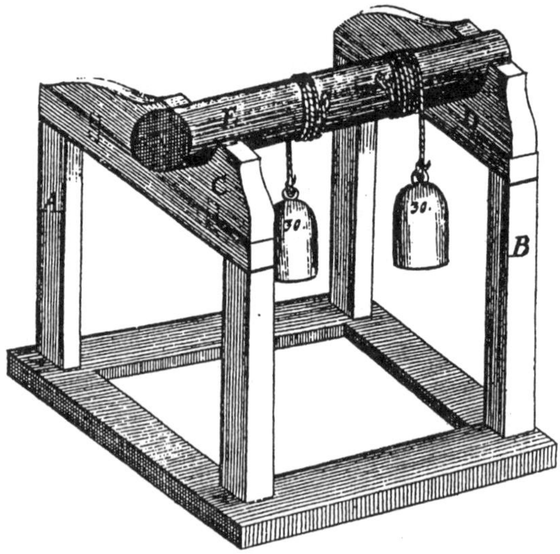

Figure 2 Leupold's apparatus for investigating the friction of bearings.

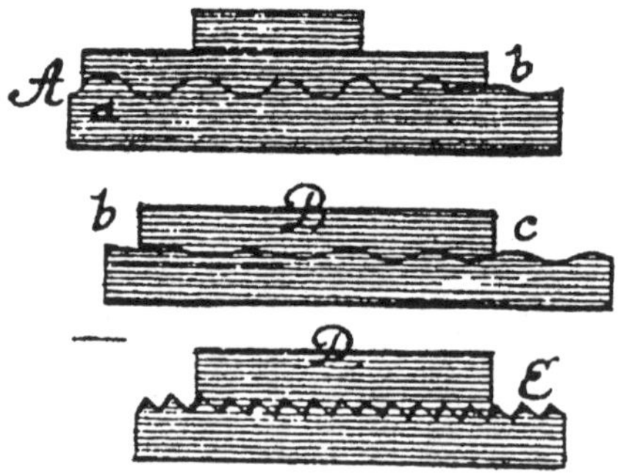

Figure 3 Different forms of roughness.

surface are very steep as in the line ab, [Figure 3], there must be a greater degree of resistance than in Figure B where the eminences are much flatter. For, as a weight is more easily drawn over a small flat eminence or inclined plane than over a high steep hill, so too is the case of surfaces B and A. So that ... [w]e must take into account the shape and nature of the surfaces of the cavities and eminences, for a hole or cavity of 4 inches depth may not engender so much friction as one of 1 inch, viz. when the latter is narrow or very steep while the

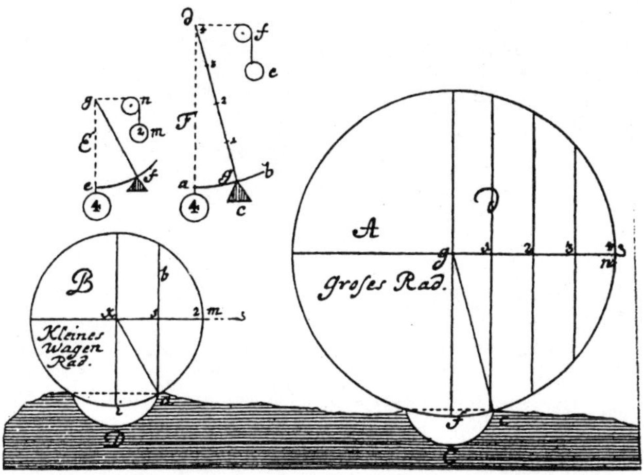

Figure 4 Small and large wagon wheels rolling over potholes.

hole of 4 inches is very wide and flat. Thus the roundness of the wheel A [Figure 4] will be drawn out of the hole C much more easily than the wheel B out of the hole D even though both holes are of the same depth and width. Therefore the higher and steeper the rough parts of the surface are with respect to each other the greater the friction, especially when very steep as [Figure 3] at D. Here the prominent parts are admittedly not high, yet, on account of their almost perpendicular frontage, cannot so well be separated or overcome in the horizontal direction when D is drawn against E, unless such points are in due course blunted or worn away by the violence applied. For which, depending on the nature of the materials, up to 20 times more force is required than for the weight, or, alternatively, the line of motion might at the same time be so directed as to help raise the weight out of the cavity.

The small diagrams adjacent to Figure 4 are particularly noteworthy in that they show Leupold's approach to a 'resolution of forces' type of analysis in the case of pulling a wheel over an obstruction. The above account not only substantiates the variability of μ, but also shows that the force required to move an object may vary depending on its direction relative to the plane of contact (an aspect further illustrated in Figures 5 and 6). These diagrams indicate, too, that Leupold had some understanding, albeit intuitive and unarticulated, of the concept of the *angle of friction*, ø ($= \tan^{-1} \mu$).

Regarding the friction in gearing, Leupold uses the somewhat unclear Figures 7 and 8 to illustrate how the mutual angle at the point of contact has an effect on the frictional resistance. A similar point is made later (§234) in considering the operation of stamping-mills (Figures 9 to 11). Here Leupold's own method of reducing the friction is to employ a cam, bearing on a roller attached to the stamp-rod, which is itself in roller guides.

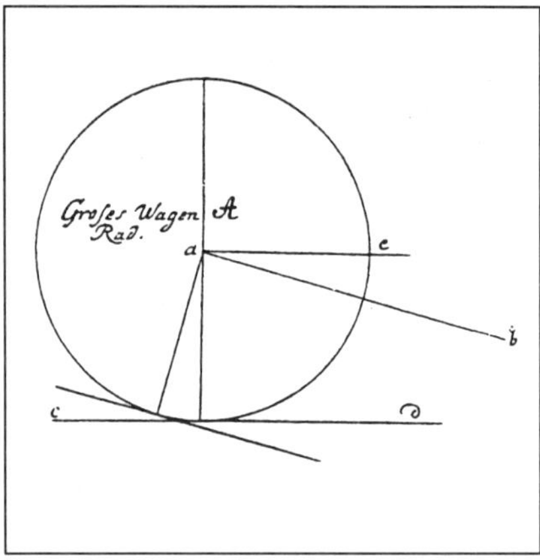

Figure 5 Wheel with inclined tractive force.

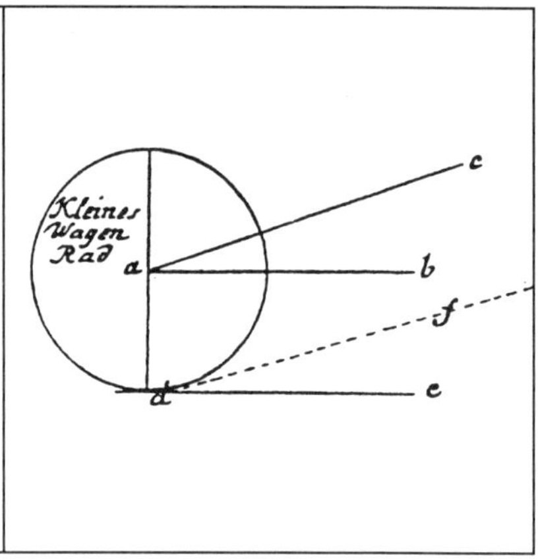

Figure 6 Wheel with inclined tractive force.

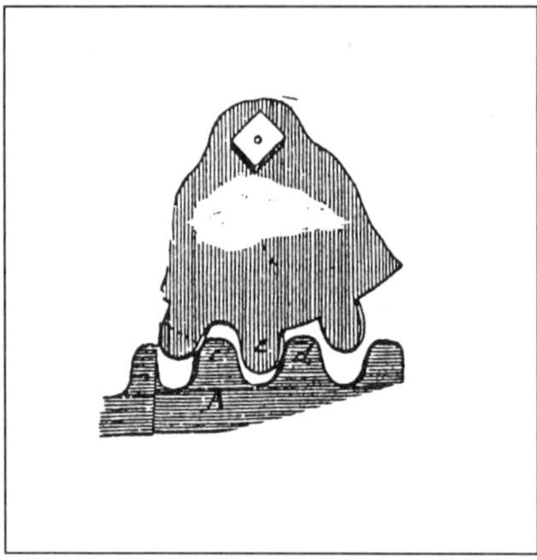

Figure 7 Friction of gear teeth.

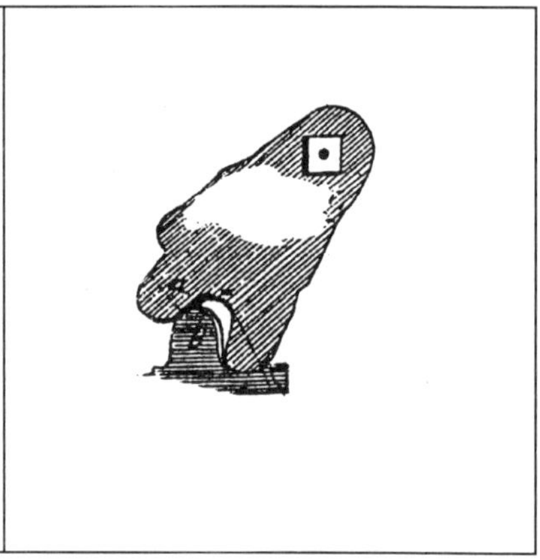

Figure 8 Friction of gear teeth.

In §223, Leupold notes that under certain circumstances the area of contact can indeed have an effect on the frictional resistance encountered:

> The rule regarding the area of surfaces suffers a considerable exception when the rough parts (as in [Figure 3]) are broken down by the force exerted. It is easy to conceive that a larger surface will have a greater number of such protuberances, thus requiring a greater force to initiate movement. But this no longer holds once movement has been achieved since this breaking-down action has only to occur across the width of the moving part. . . . Thus in a sledge it is preferable for the runners to be long and narrow rather than short and wide.

Here Leupold's perception of what occurs when one body slides on another represents an effort to explain not only the possible influence of the surface

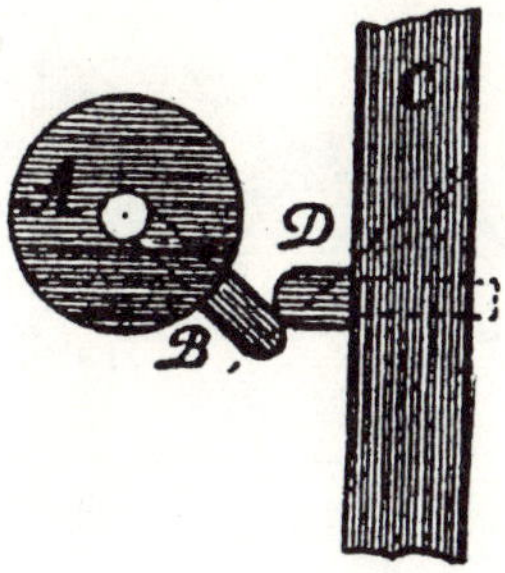

Figure 9 Stamp-mills: simple trip.

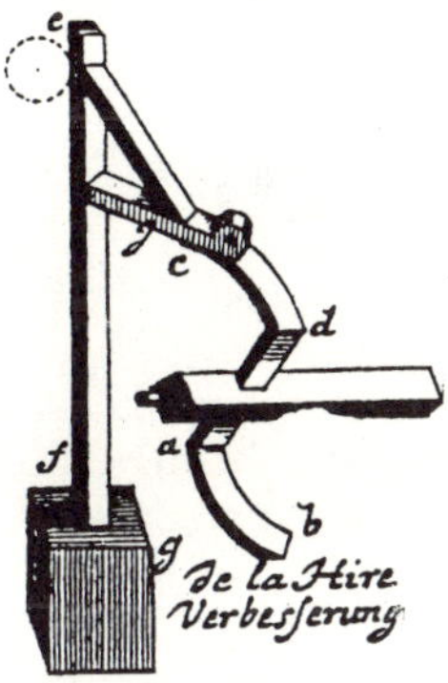

Figure 10 Stamp-mills: La Hire's cam.

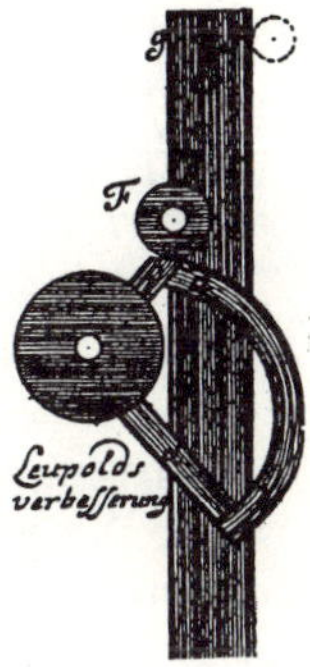

Figure 11 Stamp-mills: Leupold's cam.

area on friction, but also the observed difference between 'kinetic' and 'static' friction (the force needed to *sustain* motion being less than the force needed to *initiate* motion).

In §227 Leupold refers again to Figure 4 to show that in the case of wheels the frictional resistance occurs not only in the axle bearings but also in the rolling of the wheel along a rough surface. He notes that, while (*ceteris paribus*) the larger the wheel the less will be the 'rolling friction', the resistance in the bearing will depend on the ratio of the axle diameter to the wheel diameter (essentially his rebuttal of Sturm's criticism of Amontons's conclusions).

Friction is involved in the common arrangement of reciprocating parts (such as pump pistons) driven by cranks and connecting rods. Some of the overall friction arises from the lateral component of the force in the connecting rod acting against the side of the cylinder. He describes the 'improvement' proposed by Sturm, which involves the rigid attachment of the connecting rod to a crosshead moving in guides (Figure 12). The reciprocating motion is achieved by the crank, as it revolves, sliding to and

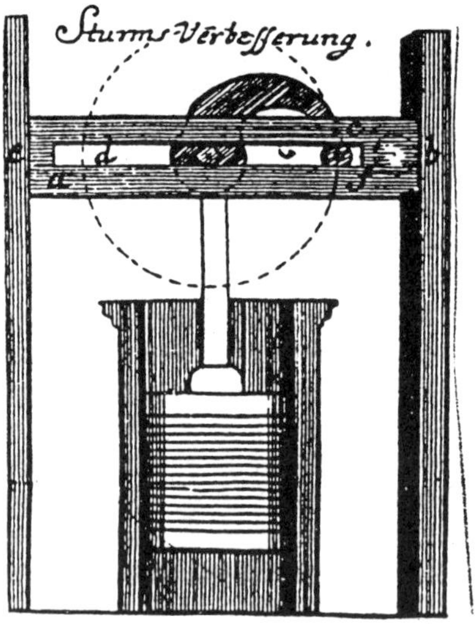

Figure 12 Sturm's crosshead.

fro within a slot in the crosshead. Leupold points out that this arrangement still involves friction arising from a lateral force, and although this could be reduced by doubling the assembly in such a way as to cause the lateral forces to balance each other (which he further discusses below), he questions whether the friction in the gearing needed would not offset any of the advantage gained. He concludes, 'I would rather maintain that a better effect will be achieved with the simple, unembellished crank, somewhat lengthened [Figure 13], or supplemented with a roller at its piston end (as at "e" [Figure 14]).'[14] This comment reflects Leupold's dislike of unnecessary complexity.[15] However, he returns later (§248) to the crosshead concept, reproducing in Figure 15 Sturm's illustration of a double crank arrangement. Here the intention is that the two cranks will revolve in such a way that they move along the same lateral line but in opposite directions, whereby the lateral forces cancel each other while the crosshead is maintained in its proper attitude perpendicular to the guides. Leupold endorses this device, 'as long as it is not disadvantaged by the cogs and lantern', but fails completely to notice Sturm's kinematic absurdity: in order to achieve the essential effect, the two cranks must themselves revolve in opposite directions, and at one point achieve the positions shown in the lower half (plan view) of Figure 15, which is impossible when they are linked by the gearing shown.

Leupold discusses the reduction of friction by various means including the use of roller bearings, the appropriate choice of materials for the surfaces in contact, and lubrication. Evidently many of the methods and devices proposed owe more to experience and trial-and-error than 'first principles'.

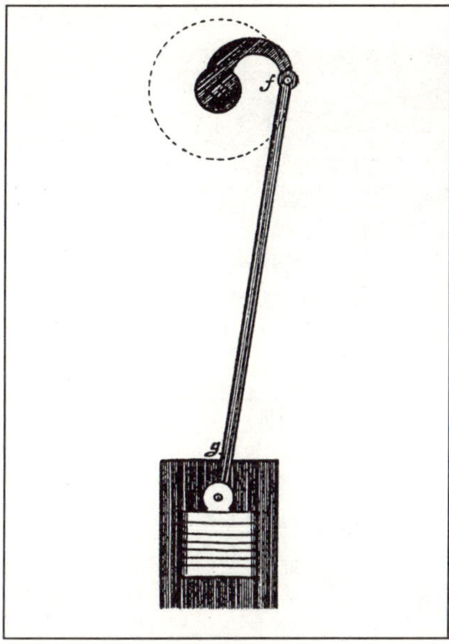

Figure 13 Crank-driven piston.

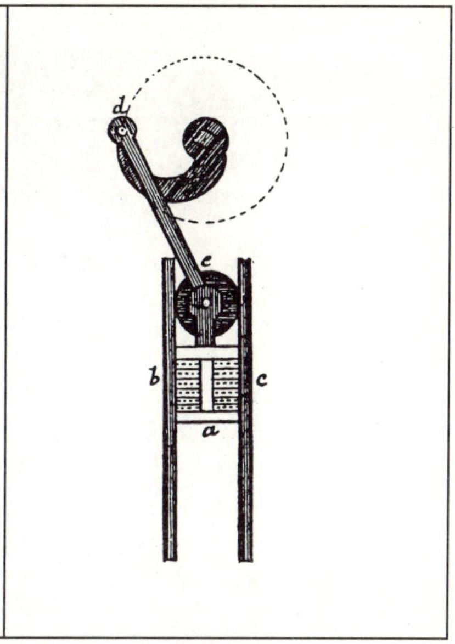

Figure 14 Crank-driven piston.

Roller bearings are shown in Figures 16 and 17. The principles and operation are self-evident. 'Rolling friction' being less than 'sliding friction', there is a considerable reduction in the torsional resistance of a bearing if the moment due to the sliding friction is relatively diminished by making the radius of the axle support greater than that of the cylindrical sliding surface. Leupold adds (§232):

> Regarding these rollers it should be noted:
> 1. that both the axles and shafts must be truly round and smooth;
> 2. that the axle should be of good iron or steel, well-fitted and hardened. For the harder and smoother, the less impression can one material make on another;
> 3. that the material for the bed [of the bearing] should be softer than brass – a hard wood is preferable to iron;
> 4. that the contact of the rollers on the bearing should be quite broad; for the broader the better in order to reduce the pressure and abrasion; similarly the axles ...
> 5. that the bearings, whether they be wood or brass, should have a groove on both sides to receive oil or grease ...
> 6. that the whole assembly be covered as a protection against dust and sand...

At the end of §239, in which he describes some tests conducted on the hardwood roller shown in Figure 18, Leupold inserts the following additional comment on lubrication:

> I afterwards smeared some journals or cylinders with oil, but found

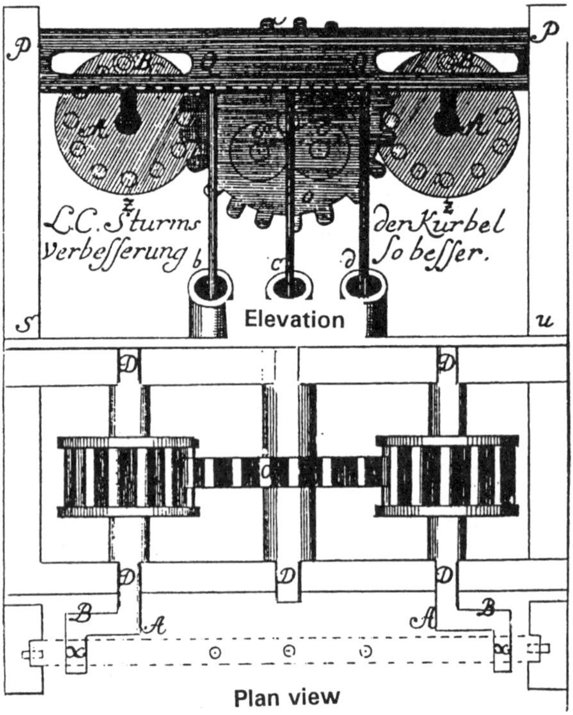

Figure 15 Sturm's crosshead with two cranks.

> that instead of ⅓ [the force needed to produce movement] I required nearly ⅔ or sometimes more. But when the oil was mixed with soap and tallow I required about 2 lb less than ⅓.

Here Leupold's powers of explanation break down, for he has no means of analysing the viscous resistance of different types of lubricant.

Leupold ends his chapter on friction by reverting to the problem of a force acting obliquely to the 'line of direction'. Once again he takes as one of his examples a cartwheel encountering a stone (Figures 19 and 20). To draw the cart forward the horizontal tractive force has to lift the load on the axle by using the structure of the wheel as a cranked lever (adc in Figure 19).

In the final chapter of the *Theatrum Generale* (following his lengthy consideration of 'the external forces in mechanics') Leupold attempts a comprehensive 'case study': the performance of a waterwheel-driven pumping installation for a mine in Freiberg, the details of which he obtained from the superintendent in charge. The case makes interesting reading, because no sooner has he started his analysis than he runs up against the familiar problem of 'inadequate data'. However, evidently because of his prime purpose to illustrate to his non-academic readers the

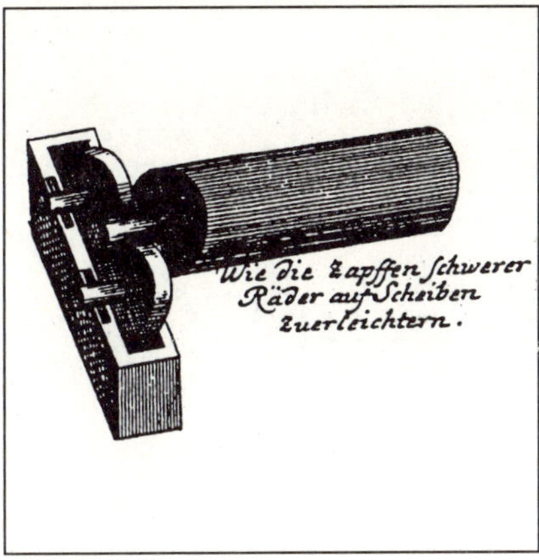

Figure 16 Roller bearings.

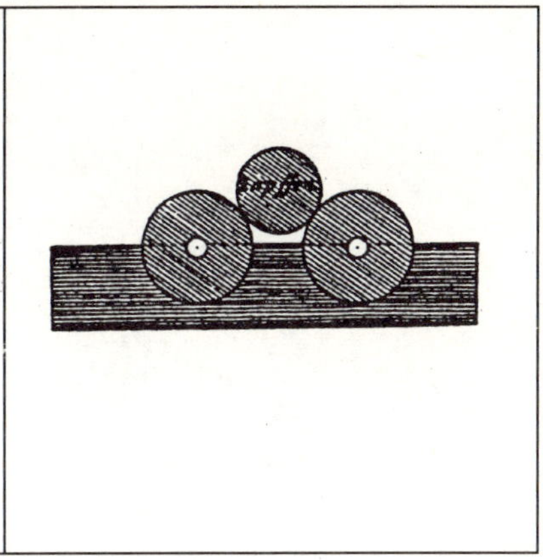

Figure 17 Roller bearings.

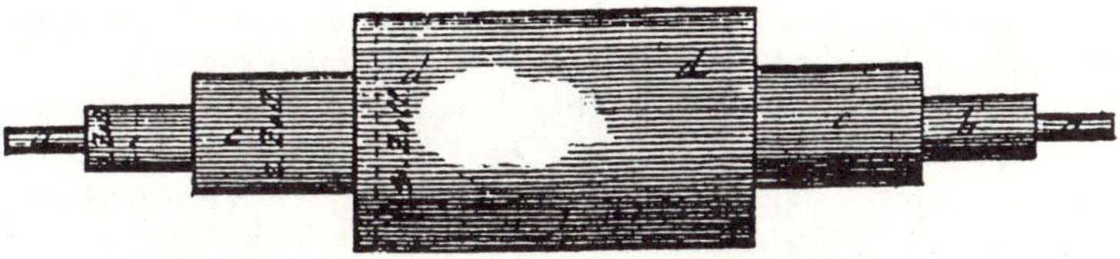

Figure 18 Leupold's experimental roller.

practical application of the first principles of mechanics, he presses on and quite openly fills in the gaps by guesswork. His analysis includes consideration of the friction in the machinery, which he feels is even more difficult to calculate than the power of the wheel. Referring to his Chapter XVI, he says (in effect) that he will assume a coefficient of friction of $\frac{1}{3}$, but 'since we lack the true weight of the wheel we will again fail to arrive at a correct answer'. None the less he lists the following data needed for the calculation of mechanical friction in the installation:

1. the weight of the wheel;
2. the weight of the water on the wheel;
3. the radius of the wheel;
4. the diameter of the axle;
5. the load on the wheel and the friction on the cranks and connecting rods; and
6. the friction of the pistons.

With his assumed friction coefficient and by candidly adopting arbitrary values for the unknowns Leupold manages to arrive at an estimate of the total frictional resistance, which he expresses as an additional force to be applied at the rim of the waterwheel. He remarks on these calculations:

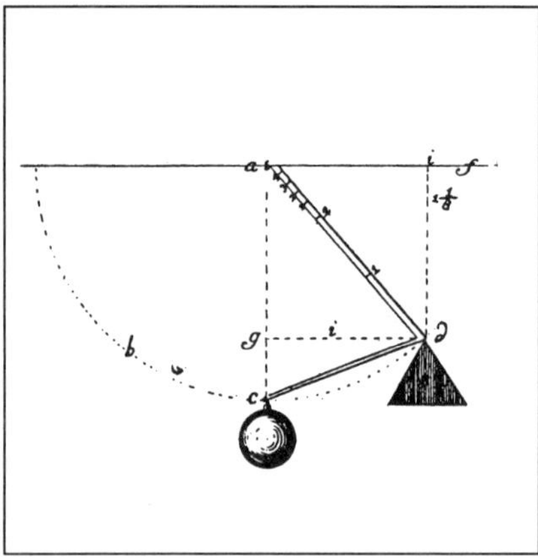

Figure 19 Oblique force acting on wagon wheel.

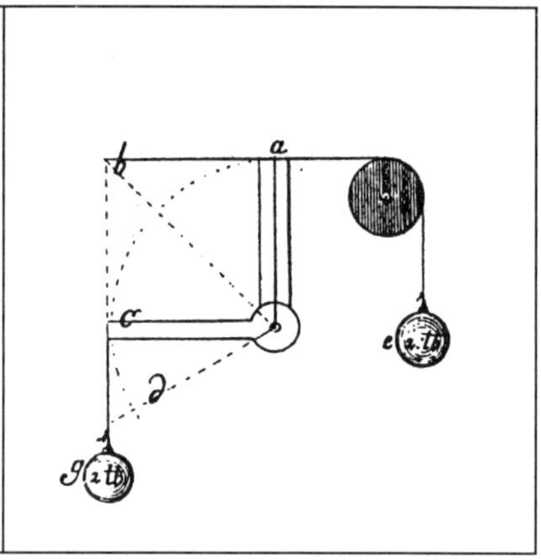

Figure 20 Oblique force acting on wagon wheel.

> We must note, however, that the [coefficient of friction of] ⅓ does not apply to lubricated journals, but rather to wooden ones coming straight from the turners, and the friction with a well-lubricated iron journal has yet to be determined. Thus all efforts to arrive at a correct result must prove fruitless without proper tests, which however require time, expense, mathematical and scientific knowledge, and someone to examine everything with an unrelenting diligence. But nothing will be done until Princes themselves encourage and promote such an undertaking, it being too expensive for private individuals.

It will be seen from the above review that Leupold's treatment of friction is limited in that he takes as his model the dragging (or rolling) of a load over 'macro' roughnesses such as protuberances on a surface. In this he is merely following the lead of Amontons, albeit with more and better illustrations and with considerably more reference to practical applications. It is noteworthy that he is aware not only of the need for systematic research into the nature of friction and the means of reducing it, but also that such research should be the responsibility of the state and not left to individuals.

This immediately invites comparison with the situation in France, with its nationally recognized system of 'college education' for engineers, where in 1781 the 'classical' work on friction was published by Charles Augustin Coulomb.[16] Coulomb still took Amontons's work as his starting point, and, although he makes no reference to Leupold, his model (of macro-protuberances) differs little from that described by Leupold. However, he also had the facilities to conduct numerous additional experiments and developed expressions which took into account elements ignored by Amontons, such as the length of time the surfaces in contact have been in a state of repose, the inclination of the sliding surfaces and, significantly, a factor embodying the area of contact. The latter case can be written in the form:

$$\text{Force (required to move load } P) = A + \mu P$$

If *A* is taken as the product of the area of contact and the limiting sheer stress between the surfaces, this equation is still in common use where resistance to movement is a combination of simple friction and 'cohesion'. It was most probably the added cohesive element which caused the increased resistance noted by Leupold in the case referred to above.

That Coulomb derived no inspiration from Leupold's essay should come as no surprise. He had entered the engineering profession by the well-established French procedure of attending a recognized college (Mezières), and later pursued a parallel career in physics research. He was indeed (in Leupold's terms) as much a *mathematicus* as a *mechanicus*. Moreover, 'scholarly works' of the early eighteenth century, if not written in French, still tended to be in Latin. Coulomb would surely have found Leupold's text, written in the vernacular for an audience of artisans, tediously repetitive and altogether too simplistic in approach.[17]

What, then, is the significance of Leupold's work, as exemplified by the isolated example of friction in machines? It would be difficult to claim that he advanced the theory of friction to any significant degree, but it is quite clear that he fully grasped the principles enunciated by Amontons, and his refutation of Sturm's argument was noteworthy. He took the pains to replicate Amontons's experiments and carried out further tests of his own, and, most importantly, he presented the prevailing state of knowledge to his intended audience of practical artisans in a lucid manner, without sacrificing the elementary 'science' on which this knowledge was based. While asserting that engineering should be rooted in the mathematics of classical mechanics (the rational approach), he clearly recognized the need for experimentation – which forms the basis of so many of the empirical formulae still in use by engineers. This philosophy, in a country such as Saxony yet to adopt the rigorous system of technical education already being developed in France, was the basis of Leupold's appeal to the practising craftsman to re-examine critically his traditional rules of thumb.

Notes and References

1. Edwin T. Layton, Jr, 'Mirror-Image Twins: The Communities of Science and Technology in 19th-century America', *Technology and Culture*, 1971, 12: 562–80.

2. The anonymous obituary published in *Neue Zeitung von gelehrten Sachen*, Dresden, April 1727: pp. 339–43. A brief announcement of Leupold's death appeared in the January 1727 issue, 63–4.

3. He can also be compared with John Smeaton (1724–92), whose initial work as an instrument-maker brought him in contact with prominent members of the Royal Society and led to an eminent career as a consulting engineer.

4. Ulrich Troitzsch, 'Zum Stande der Forschung über Jacob Leupold', *Technikgeschichte* 1975, 42(4): 263–86.

5. Jacob Leupold,
Theatrum machinarum generale (Leipzig, 1724)
Theatrum machinarum hydrotechnicarum (Leipzig, 1724)
Theatri machinarum hydraulicarum: Tomus I (Leipzig, 1724)
Theatri machinarum hydraulicarum: Tomus II (Leipzig, 1725)
Theatrum machinarum oder Schauplatz der Hebzeuge (Leipzig, 1725)
Theatrum staticum universale (Leipzig, 1725)
Theatrum pontificiale (Leipzig, 1726)

Posthumous:
Theatrum arithmetico geometricum (Leipzig, 1727)
Theatrum machinarum molarium (Leipzig, 1735)
Theatrum machinarum supplementum (Leipzig, 1739)

NB. The text in each of the volumes of the *Theatrum Machinarum* is divided into chapters, with a parallel subdivision into paragraphs, which are usually numbered (e.g. §415) continuously from the start of each volume. The illustrations in the *Theatrum* are on *tabulae* with several figures on each *tabula*. The numbering of both the *tabulae* and (with few exceptions) the figures is in Roman numerals.
References to the original *Theatrum* will be made by volume (only when required by the context) and paragraph numbers. Reproduced figures have been renumbered.

6. *T. M. Generale*, Preface, 4th page (the pages and paragraphs of the Preface are unnumbered).

7. A.G. Keller, *A Theatre of Machines* (New York: Macmillan, 1964). The examples mentioned are Agostino Ramelli, *Le diverse et artifisiose machine* (Paris, 1588); Jacques Besson, *Théâtre des instruments mathématiques et mécaniques* (Geneva, 1594); Fausto Veranzio, *Machinae Novae* (Venice, *c.* 1615); and Vittorio Zonca, *Novo Teatro di Machine et Edificii* (Padua, 1656).

8. A typical example would be the case of flow measurement, where Leupold refers to Edmé Mariotte, *Traité de mouvement des eaux et des autres corps fluides* (Paris, 1686). See also translation by J.T. Desaguliers, *The Motion of Water and Other Fluids* (London, 1718).

9. Leupold's figures are often no more than untitled thumbnail sketches with cryptic legends, evidence that their sole purpose is to support the arguments in his text.

10. Troitzsch, *op. cit.* (4).

11. It has been said of Ramelli's book (see note 7 above) that 'Quelques-unes de ces machines sont assez ingénieuses; mais elles seraient plus utiles si elles étaient plus simples' (Some of those machines are rather ingenious; but they would be more useful if they were simpler) (*Biographie Universelle*).

12. G. Amontons, 'De la résistance causée dans les machines', in Académie Royale des Sciences, *Mémoires* for 1699 (Paris, 1732). Generally, throughout the *Theatrum* Leupold's citations are sufficiently meticulous for his sources to be readily traced.

13. L.C. Sturm, 'Observationes circa frictionem', in Akademie der Wissenschaften, *Miscellanea Berolinensia*, 1710, 1(3): 294–307.

14. *Theatrum Generale*, §238. It may be noted *en passant* that both Figures 13 and 14 show 'traditional' cranks in which the arm has a pronounced curve. Leupold has already (§148) demonstrated the uselessness of this arrangement in his discussion of the crank *per se*.

15. The *Theatrum Generale* ends with a list of factors to be taken into account in the design of machines (§631), including the following: 'The simpler the machine, the better the efficiency and the less the cost.'

16. Charles Augustin de Coulomb, *Théorie des machines simples* (Paris, 1821) (a collection of scientific papers by Coulomb). For a comprehensive account of Coulomb's achievements see C. Stewart Gillmor, *Coulomb and the Evolution of Physics and Engineering in Eighteenth-Century France* (Princeton, 1971).

17. And we must also recollect that in the preparation of *l'Encyclopédie* Diderot had reviewed the *Theatrum* but considered it 'inutilisable'. See Jacques Proust, *Diderot et l'Encyclopédie* (Paris, 1967), p. 178.

Steam and Sugar: The Diffusion of the Stationary Steam Engine to the Caribbean Sugar Industry 1770–1840

JENNIFER TANN

INTRODUCTION

The stationary steam engine is a frequently cited symbol of industrialization in Western Europe and North America,[1] and, while steam engines were sought by some rulers and aristocrats in the underdeveloped world partly as a status symbol,[2] the more rapid diffusion of steam power took place in the industrializing world in association with the growth of the extractive and manufacturing industries.[3] There was one major exception. The diffusion of steam power to the plantation economy of the Caribbean was on a scale that far exceeded any other overseas market for British engines, the demand being almost exclusively for engines for sugar cane rolling.

The Caribbean sugar industry has received attention from a number of scholars, in particular from the perspectives of slavery and the plantation economy.[4] The technology of sugar milling has received less attention.[5] Exceptions include Deerr, whose two-volume study *The History of Sugar* was published in 1949–50, and Watts, who devotes part of his major book *The West Indies: Patterns of Development, Culture and Environmental Change since 1492*, to the study of innovation diffusion.[6] It is now timely to reopen the discussion not only to present new data on the application of steam power to sugar milling, but to establish a contextual framework within which to locate the evidence.

Within the developing British engineering industry two firms, Boulton & Watt of Birmingham and Fawcett & Littledale of Liverpool, dominated the early-nineteenth-century steam engine export market, and for both of them the Caribbean comprised a major element of their overseas sales.[7] Watts states that 'steam was utilised only spasmodically prior to emancipation',[8] yet in aggregate Boulton & Watt and Fawcett &

Littledale supplied 267 stationary engines to the Caribbean between 1803 and 1825 and a further 48 between 1826 and emancipation in 1833. Of the engines ordered up to and including 1825 Boulton & Watt sold 119, a figure which exceeded the rest of its sales overseas (Ireland excluded) in the fifty-year period from the commencement of their engine business to 1825. Between 1778 (the year of its first overseas sale) and 1825 the company sold 110 engines to other overseas markets including 46 to Western Europe, 15 to Southern Europe, 11 to East and Central Europe, 15 to North America and 23 to the non-industrialized world including 15 to India.[9] Moreover in terms of numbers of engines utilized by a single industrial sector, Boulton & Watt's sales to the Caribbean were exceeded in the British domestic market during the same period only by engines supplied to the cotton industry,[10] the largest industrial user of the Watt engine to 1825. Fawcett & Littledale supplied 148 engines to the Caribbean in the period to 1825. Unlike Boulton & Watt, whose market was characterized by diversity, Fawcett & Littledale, reflecting its local business environment, targeted two user sectors almost exclusively: sugar and shipping. This paper will consider the diffusion of steam engines to the Caribbean, addressing the nature and level of demand for motive power, the choices available for meeting the demand, and early precedents for steam power in the Caribbean in the light of theories of colonial economic development and international technological diffusion.

THE EVOLUTION OF THE SUGAR MILL

The successful sugar planter had to be a skilled manufacturer as well as an agriculturalist, a technician as well as a businessman. Thus 'the effective use of the mill is always a key factor in the manufacture of high quality sugar'.[11] The design of the Caribbean sugar mill was diffused from the Canary Islands to Española (Hispaniola) in the early sixteenth century, being based on the two-roll design of Pietro Speciale of Palermo, Sicily.[12] The origin of the three (vertical) roller mill is unclear but it was employed in the Pernambuco region of Brazil in the mid-seventeenth century and introduced to the West Indies by James Drax. With this design, which remained in use until the nineteenth century, the cane was passed between the rolls twice, two slaves standing either side of the rolls, passing the cane through first one way and then the other. The requirement for a second slave was obviated by the invention, *c.* 1755, of the 'double-use' or 'dumb' returner, an iron shield that caught the once-crushed cane and oriented it for automatic re-entry to the rollers for the second crushing. Like a number of other incremental improvements to the sugar-mill, this was not widely implemented.[13] Experiments in reorienting the rolls to the horizontal rather than the vertical plane resulted in the triangular arrangements introduced at the end of the eighteenth century in which the third roller was placed above two laid horizontally, juice extraction being aided by fluted rollers. This design proved to be readily adaptable to steam power but the vertical pattern continued in use until well after emancipation.[14]

NATURAL SOURCES OF POWER

The adoption of one natural source of power in preference to another depended, in part, on environmental factors: terrain, rainfall, the regularity of prevailing winds, incidence of hurricanes, the availability of animal feedstuffs, and the availability and price of construction materials. In addition there were comparative fixed and recurrent costs to consider, besides the size of plantation and the milling capacity required. Each source had disadvantages. Windmills were effective only along windward coasts, exposed hill sites and areas fully cleared of forest. A lull in the trade winds during the dry, cane-harvesting season could lead to the temporary cessation of milling with a likelihood of heavy losses from the postponed milling of cut cane and the delayed and more difficult transport of sugar to the nearest port during the rainy season. Windmills predominated on Barbados and Guadeloupe, and comprised more than one-quarter of the power systems on Tobago and Montserrat. Some of the windmills on Barbados were 30 ft or more in height, and attention has been drawn to similarities in design to those of Britain and north-west France.[15] Whereas some estates in Barbados had two mills and a very few had three in the early eighteenth century, the number of multiple mill estates declined after adverse economic conditions in the 1730s. Poor trading conditions in the early nineteenth century caused a reduction in the number of windmills on Barbados, 335 being recorded in 1813.[16] Watermills were much less frequently employed on West Indian sugar plantations, usually being located on the larger, more mountainous islands. In general, the absence of suitable rivers deterred the wider adoption of water power, but it is noticeable that water power was most effectively implemented where French engineers were employed, as in St Domingue (Hispaniola) in the 1740s and 1750s. Numerous watermills were constructed in Guadeloupe and Martinique (Table 1), the French influence also being apparent in Grenada, where there were 95 watermills only a few years after its transfer to British control. The only other West Indian island with a significant number of watermills was Jamaica, although it has been suggested that water power was less fully exploited than might have been expected.[17] Nevertheless, there were some major investments in water power engineering, the 30 ft diameter wheel of one mill being fed by a 2 mile leat in the 1680s.[18] The majority of Jamaican plantation owners who had watermills, however, also owned animal-powered mills.

Animal power was the most widely employed of the natural sources of power. Cattle, horses, mules, donkeys and, very rarely, slaves were used. Horses seem to have been preferred in the seventeenth century, the price falling from £50 per animal in 1650 to £15 in the 1670s. Few were bred in the islands, most of them being imported. By the early eighteenth century there was some debate about the most effective mill animals, cattle being preferred on the French islands. A disease which killed off many of the horses in Barbados in 1714–18 led to cattle being preferred thereafter. Mules were employed in Jamaica, and in the northern Leewards horses, mules and assingoes (the Azorean donkey) were used.

Table 1 Natural sources of power in the Caribbean

Territory	*Year*	*Windmills*	*Watermills*	*Animal mills*
Barbados	1748	393	—	—
	1789	446	—	—
	1813	335	—	—
Antigua	1705	36	—	134
Martinique	1769	12	116	184
	1820	20	178	199
Dominica	1775	15	6	20
Guadeloupe	1738	1	80	174
	1767	11	144	263
	1790	140	133	228
	1818	222	136	117
Tobago	1775	23	9	52
Trinidad	1808	7	8	257
Grenada	1772	12	95	18
St Lucia	1789	3	32	18
Jamaica	1768	49	235	369
Montserrat	1729	23	3	52

Sources: N. Deerr, *The History of Sugar* (London, 1949); D. Watts, *The West Indies* (Cambridge, 1987); R.B. Sheridan, *Sugar and Slavery* (Baltimore, 1973).

In most of the islands for which early data are available, animal mills exceeded other sources of motive power (Table 1),[19] and it has been estimated that, at the end of the eighteenth century, approximately one-third of the British West Indies sugar crop passed through mule mills.[20] Although the capital cost of a watermill was the highest of the three natural sources of power in the Caribbean, followed by that of a windmill, the housing provided for mill animals at some plantations, perhaps to extend the animals' working lives, would seem to have been left out of the equation when the total fixed costs of power are compared. Moreover, the recurrent costs of animal power were higher than either of the other two natural prime movers. Animals had short working lives and had to be replaced, and feedstuff was required.

In contrast to the employment of animal power in Northern Europe (where the owner of a typical animal 'engine' would have from 2 to 10 beasts),[21] a plantation sugar-mill in the Caribbean might have between 40 and 50 mules and steers on account of shorter working 'shifts' and the high mortality rates occasioned by the climate.[22] Compared with Northern Europe, where animal power was a low-cost prime mover, in the Caribbean the total cost differential between animal power and other prime movers was less clear, and may even have been inverted. There was thus a considerable incentive to find a substitute for animal power, even though it was only for seasonal use.[23]

EARLY STEAM ENGINES

The potential market for steam power on sugar plantations was recognized in two patents granted in the mid-eighteenth century, in one of which the patentee focused on a prime mover for sugar milling, an unusually exact specification for a steam engine at that time.[24] The first patent (no. 859) was granted in 1766 to John Stewart for an invention 'to work mills ... [which] is particularly useful and profitable in grinding sugar canes' (Figures 1 and 2). He claimed that his engine would 'grind more than a windmill and a cattle mill usually do'[25] and 'is most profitable to grind sugar canes as one of them in which a cylinder of 30 inches diameter is provided will do as much work as four cattle mills that employ 160 or 200 steers'.[26] This invention, Stewart claimed, could save up to £600 per annum on every estate employing eight or ten sugar boilers, in addition to £1200 to £1500, the initial cost of building a mill and purchasing cattle. The second patentee was Dugald Clark, whose patent (no. 949) was granted in 1770.[27] Although many a patentee may have been tempted to claim rather more for an invention than was proven in fact, one of Stewart's engines had been erected on Jamaica by 1769, receiving the endorsement of the Jamaica Assembly: 'There is the greatest prospect of it [the engine] answering the purpose for which it was intended as the power of the said engine is found to be sufficient for the grinding of the canes'.[28] A committee of the Assembly proposed that a Bill be brought in 'to give the petitioner the sole right of erecting such mills for a certain time', and an Act was passed in December 1770 enabling Stewart to 'carry into execution his newly invented mill for the grinding of sugar canes with the power of a fire engine'.[29] But after the engine's installation Stewart encountered difficulties. From the outset there was insufficient water for the boiler because the island was in the grip of a drought. And then the engine met with carefully orchestrated local resistance on two fronts. First, a local millwright, the engine patentee Dugald Clark, tried to sell a pumping engine and waterwheel to the plantation owner on whose estate Stewart's engine had been erected and then installed two alternative mills in an endeavour to demonstrate that Stewart's mill would not work. Second, the estate overseer appears to have done what he could to deprive Stewart's engine of water and fuel and the mill of sugar cane, asserting that it consumed more fuel than any estate could supply and stopped so frequently that it was unable to do as much work as two cattle mills.[30] Thus the experiment ended.

Interest in the application of the Boulton & Watt engine to sugar milling also dates from the 1770s. William Pultney[31] wrote to the partners in 1776, referring to the work of both Stewart and Clark and noting that neither could make steam power succeed, adding 'I can hardly conceive any thing which is more likely to bring you a great return.' In both 1786 and 1787 Lord Penryhn[32] drew attention to the need for steam engines in Jamaica and in 1788 George Glenny,[33] a millwright, wrote from Jamaica advising that a 16 inch cylinder engine would do the work of more than two cattle mills. Two years later John Dawson of Liverpool inquired after

A

DESCRIPTION

OF A

Machine or Invention

To Work MILLS,

By the Power of a FIRE-ENGINE,

But particularly uſeful and profitable in Grinding SUGAR-CANES,

As it will Grind more than a Wind-Mill, and a Cattle Mill uſually do, without any Expence of Fuel, more than is conſumed in boiling the Sugars. By which it appears, there may be ſaved upwards of £600 Yearly, in every Eſtate which employs Eight or Ten Coppers; beſidemany other Advantages.

And in New Eſtates may be ſaved from £1200, to £1500, in the firſt Expence of Building the Mills, and purchaſing the Cattle, &c. Beſide many other Advantages.

Which ENGINE, MACHINE, and SUGAR-MILL, may be ſet up to Work in Twelve or Fifteen Days Time.

TO WHICH IS ANNEXED

A Plan of a Fire Engine, the ſaid Machine, and a Sugar Mill and a Boiling Houſe.

With an APPENDIX.

Which ſhews, in a few PAGES,

The Utility of the ſaid Invention, and Anſwers and Explains every Objection or Doubt of it anſwering the Deſign which the Writer has heard of.

Alſo a calculation of the Power of Cattle in Grinding Canes, and the Power of FIRE ENGINES in the ſame, with References to the ſaid Deſcription.

By the King's Patent to JOHN STEWART.

Figure 1 Title page of John Stewart's pamphlet of 1767.

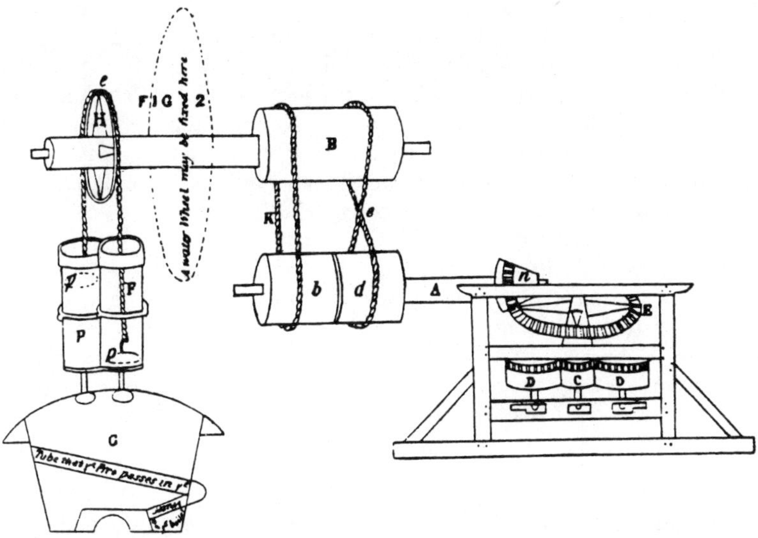

Figure 2 Dugald Clark's patent. Elevation.

an engine for a Trinidad mill, as 'the want of wind and water, the principle on which they are at present worked, retards the progress so very much, particularly at crop time'.[34] It was in 1789 that a clear connection was drawn between the anti-slavery movement and steam engines, when Samuel Whitbread[35] raised with Boulton & Watt the question of the employment of steam power in the West Indies. He reported, 'Mr Wilberforce has inquired of me about [them] and I think you told me some West Indian planter has consulted you on them and that you gave some reasons why you thought they would not answer him.' In the following year Whitbread[36] attempted to provide further information through an intermediary, later in the same year proposing[37] a subscription scheme to which Lord Penrhyn intended being a subscriber, the underlying issue being a belief that the diffusion of steam power to the West Indies might contribute to the abolition of slavery. But sugar-mills employed a small proportion of the total plantation slave labour. An animal mill generally required five slaves. An efficient steam-powered mill might operate with three. Mill capacity did, however, circumscribe plantation size, or at least the amount under cane. As the price of slaves increased from the late eighteenth century, steam power was one innovation that could, in principle, contribute to modest savings in recurrent costs, besides effecting a small reduction in the dependence on slaves.

In 1794 the Chevalier Betancourt[38] applied to Boulton & Watt for two engines on behalf of plantation owners in Cuba. Betancourt was an engineering officer in the service of Spain; it was thought likely by Boulton that the friends on whose behalf Betancourt inquired were the Marquis de

Casa Montalro and Don Fransisco de Arango. Arango, who visited England, Portugal, Barbados and Jamaica in 1795, dispatched detailed information to Cuba on sugar-mill technology, including refining methods, in the belief that 'those marvellous European machines'[39] would be of great advantage to the Cuban economy. It was Boulton[40] who first urged that action be taken in regard to the Caribbean market with specific reference to Betancourt: 'If we can make ourselves safe I think we would not drive him into the hands of Trumpeton Founders or Bull[41] or any men who are likely to bring engines into disgrace in the West Indies.' No order was placed with Boulton & Watt by Arango but the earliest successful application of steam power to cane crushing is likely to have been the imported engine of unknown (but almost certainly British) provenance erected in Cuba in 1797.[42] Although it was to be nine years before a Boulton & Watt engine was ordered for the Caribbean, Boulton's desire to 'make ourselves safe' was realized, for the majority of Boulton & Watt's engines destined for the Caribbean were ordered through mercantile houses of some repute.[43]

BOULTON & WATT AND FAWCETT & LITTLEDALE

The first Boulton & Watt engine destined for the Caribbean was ordered in 1803, but despite the early interest in steam power in the region, orders were slow until after 1810 (Table 2). From 1812 to 1816 they more than doubled, falling back thereafter to fewer than ten per year, and after 1820, to fewer than five. In all, 119 Boulton & Watt engines were ordered between 1803 and 1825. Thereafter the orders ranged from zero to four engines per year, a total of 25 being ordered between 1826 and 1852 (Table 3).[44] During the period 1803–25, the destination of the majority of Boulton & Watt engines was Jamaica (35.3 per cent of the total to the Caribbean) or Guiana (42.9 per cent), the majority of the latter going to Berbice. Other engines went to Trinidad, Grenada, Tobago, St Lucia, St Croix and Surinam (Table 4).

Fawcett & Littledale's surviving order books,[45] which commence in 1813, show what appears to be an already established Caribbean trade, twelve sugar-mill engines being ordered in that year. With the exception of 1815, when only five engines were ordered, the annual figure was over ten and on three occasions over twenty between 1813 and 1825 (Table 5). In all the firm dispatched 148 sugar-mill engines to the Caribbean between these years. Exports were maintained at a higher level than for Boulton & Watt engines from 1826 onwards, 101 engines being ordered between 1826 and 1845 (Table 6). The destinations of Fawcett & Littledale engines in part corresponded with and in part complemented those of Boulton & Watt engines. They included Guiana (93 engines between 1813 and 1825) and Jamaica (9 engines) (Table 7). The market in which Fawcett & Littledale had a unique presence was Cuba. Between 1813 and 1825 the firm dispatched 20 engines to Cuba and from 1826 to 1845 a further 54 (Table 8).

Table 2 Boulton & Watt: engines ordered for the Caribbean 1803–25

Year	*Number*
1803	1
1804	2
1805	4
1806	0
1807	0
1808	4
1809	2
1810	3
1811	5
1812	8
1813	15
1814	14
1815	19
1816	13
1817	9
1818	7
1819	1
1820	5
1821	1
1822	0
1823	4
1824	1
1825	1
Total	**119**

Source: BRL, Engine Books, Catalogue of Old Engines, letters.

Table 3 Boulton & Watt: engines ordered for the Caribbean 1826–52

Year	*Number*
1826	1
1827	1
1828	2
1829	4
1830	2
1831	0
1832	0
1833	0
1834	0
1835	1
1836	1
1837	1
1838	0
1839	1
1840	0
1841	2
1842	3
1843	0

Table 3 *(cont.)* **Boulton & Watt: engines ordered for the Caribbean 1826–52**

Year	*Number*
1844	0
1845	0
1846	3
1847	0
1848	0
1849	0
1850	1
1851	1
1852	1
Total	**25**

Source: BRL, Engine Books, Catalogue of Old Engines, letters.

Table 4 Number of Boulton & Watt engines ordered for different islands/colonies of the Caribbean as a percentage of the total distributed to the whole area 1803–25

Island/colony	*Number ordered*	*Percentage of total for Caribbean*
Guyana (Berbice, Essequibo, Demerara)	51	42.9
Jamaica	42	35.3
Trinidad	12	10.1
Grenada	4	3.4
Tobago	1	0.8
St Lucia	1	0.8
St Croix	2	1.7
Surinam	1	0.8
Unknown	5	4.2
Total	**119**	**100**

Source: BRL, Engine Books, Catalogue of Old Engines, letters.

However, although in terms of the export of British engineering goods in the early to mid-nineteenth century the number of engines dispatched to the West Indies is impressive, steam-powered mills comprised a small proportion of the total number of mills on Caribbean sugar plantations at that period. Figures for the number and size of sugar plantations in the West Indies in the early to mid-nineteenth century are unavailable, but if the number of mills reported in the mid- to late eighteenth century (Table 1) is taken as an indicator for the number of sugar plantations (estates with more than one mill probably cancelling out small estates with none), the proportion of steam to natural power sources in the early nineteenth century was small. By the mid-nineteenth century steam power was an embedded technology in the more advanced sugar-producing islands such as Cuba (70 per cent of mills being steam powered by 1861), Tobago (37 per cent of mills being steam powered) and Jamaica (22 per cent steam powered by 1850).[46]

Table 5 Fawcett & Littledale: engines ordered for the Caribbean 1813–25

Year	*Number of engines*
1813	12
1814	15
1815	5
1816	22
1817	16
1818	26
1819	23
1820	5
1821	3
1822	2
1823	7
1824	4
1825	8
Total	**148**

Source: Liverpool Maritime Record Office, F&L Papers, Engine Books.

Table 6 Fawcett & Littledale: engines ordered for the Caribbean 1826–45

Year	*Number of engines*
1826	6
1827	3
1828	6
1829	6
1830	11
1831	6
1832	0
1833	0
1834	2
1835	2
1836	8
1837	7
1838	15
1839	12
1840	9
1841	2
1842	2
1843	1
1844	1
1845	2
Total	**101**

Source: Liverpool Maritime Record Office, F&L Papers, Engine Books.

Table 7 Fawcett & Littledale: destination of Caribbean engines 1813–25

Colony	*Number of engines*
Guiana	93
Cuba	20
Jamaica	9
St Lucia	8
Martinique	6
Nevis	4
Surinam	4
Trinidad	1
Haiti	1
Antigua	1
Total	**148**

Source: Liverpool Maritime Record Office, F&L Papers, Engine Books.

Table 8 Fawcett & Littledale: destination of engines for the Caribbean 1826–45

Colony	*Number of engines*
Cuba	54
Guiana	18
French Guiana	15
Antigua	6
Jamaica	4
Trinidad	2
Martinique	2
Total	**101**

Source: Liverpool Maritime Record Office, F&L Papers, Engine Books.

DISCUSSION

There is no mono-causal explanation of the large-scale diffusion of stationary steam engines to the Caribbean between 1770 and 1840. The explanations lie in elements of the market combined with specific situational factors for certain territories which converged during the late eighteenth to mid-nineteenth centuries. These included Britain's dominance as a colonial power in the Caribbean; the price of sugar and related plantation profits; the physical properties of cane; the close professional relationship between Boulton & Watt and John Rennie; and the role of London, Bristol and Liverpool merchants and sugar-house owners as third parties.

Major contributions to the debate on the relationship between British imperialism and economic growth by Cain and Hopkins[47] and Ward,[48] and to the relationship between imperialism and technological diffusion by Headrick,[49] inform an interpretation of the diffusion of steam power to the

Caribbean in the early to mid-nineteenth century. Cain and Hopkins[50] assert that agriculture and finance, rather than industry, gave shape and momentum to Britain's overseas expansion after 1688, a view acceded to by Ward,[51] but only for the first seventy-five years of the eighteenth century. By the 1790s, Ward contends, 'the main support to Great Britain's balance of payments came from the increase of its domestic exports ... derived from real advantages in manufacturing efficiency rather than from the use of force against rivals in the classic pattern of mercantilist warfare'.[52] Although, as Ward notes, the Royal Navy's success in cutting off France, Spain and the Netherlands from their colonies 'did not by itself guarantee British exporters greatly enlarged sales',[53] there is no doubt that it considerably helped in so far as the Caribbean market for steam power was concerned. And, although Ward is right to point out that the contribution of new manufacturing sectors to the British economy should 'not be judged simply by matching the pattern of overseas marketing to the geography of conquest',[54] there is a particularly close correlation between the pattern and chronology of conquest and the diffusion of engines to the Caribbean. Headrick asserts[55] that Western technologies were exported to colonial possessions for profit, scant regard being paid to their longer-term impact on colonial economies, curiously implying that a failure to consider the impact of technology transfer was unique to the nineteenth century. He is right, however, to draw attention to the possibility of technological dependency through the retention of imported technologies and their eventual substitution by newer technologies from the country of origin, a possibility endorsed by Tann and Breckin's model[56] of the international diffusion of steam power in the industrial revolution. Colonial status served to boost diffusion but retarded the development of local engineering industries.

In the Caribbean market Boulton & Watt supplied engines almost exclusively to British possessions, although the firm's preparedness to sell to countries with which Britain was at war is well known.[57] While local factors were often clearly important and Boulton & Watt did not supply engines to all Britain's possessions in the Caribbean, the potential market for engines expanded with the growth in the number of British possessions and Britain's dominance at sea in the early nineteenth century. In many cases inquiries for engines were made within a year or so of a territory passing into British hands. For example, an order was received for an engine for Trinidad in 1803,[58] a year after ownership passed from Spain to Britain following the Treaty of Amiens; and three engines were ordered three years after part of Guiana came into British hands.[59] But whereas Tobago and St Lucia were also acquired during the Napoleonic War period, they were not important markets for engines, a clear implication being that other factors were also important in technical choice. Guiana, together with Jamaica, came to dominate the Caribbean market for Boulton & Watt engines from 1813, both being confirmed as British possessions by the Treaty of Vienna of 1814. Fawcett & Littledale's largest market was British Guiana. The company also supplied engines to

Martinique during a brief period of British rule between 1809 and 1814, as well as a number of engines to Cuba, which, although technically Spanish, had become so far as trade was concerned, practically independent from the late eighteenth century under a succession of liberal governors. Free commerce with Cuba was a reality from 1809, legalized in 1819 and confirmed in 1824, a situation exploited effectively by Fawcett & Littledale.[60] Boulton & Watt supplied one and Fawcett & Littledale supplied four engines to Surinam, a Dutch possession. British dominance of the Caribbean through colonial possessions as well as at sea furnished a captive market in which there was little competition from non-British suppliers and an absence of competition from local ones.

Cain and Hopkins[61] emphasize the relationship between British overseas expansion and the state of the British economy over time. At government policy level, they assert, 'the extension of Britain's expansion can be seen as an expression of her failure to dominate her chief competitors and especially to prevent their industrialisation'.[62] Boulton & Watt's engine business focused largely on the domestic market during the term of Watt's patent (1775–1800), overseas inquiries often being responded to without enthusiasm when order books were full. In slack years, however, the partners adopted a more proactive stance in overseas markets, efforts being made to obtain patents in Spain, France, the Netherlands and America,[63] and did make positive responses to overseas inquiries. Domestic orders were buoyant in the early nineteenth century but were depressed in 1813, 1816, 1817, 1821 and 1822. In these years, 1821 apart, the slackness was compensated by an upsurge in the number of overseas orders accepted (Table 9), demand from the West Indies being particularly high between 1813 and 1816, years of boom in sugar prices. Although it would be inadvisable to lay too great an emphasis on one firm's order books, there is some evidence to suggest that Boulton & Watt's response to overseas (including colonial) markets was a reflection of government policy in microcosm.

An examination of the relationship between sugar prices and engine orders shows the 1813–19 peak in engine sales to correspond with a contemporary boom in sugar prices. Similarly the 1838–9 peak in Fawcett & Littledale engine orders corresponds with another boom. In these years plantations were more profitable and investment more attractive.

The introduction of a new strain of sugar cane – Otheite – to the West Indies in the late eighteenth century enhanced plantation productivity since it yielded more juice and furnished more fuel.[64] The composition of the cane (69–75 per cent of it consisted of water)[65] necessitated processing at the point of production in order to minimize deterioration of the product.[66] It was essential to keep 'intact the land and factory combination', and 'because of the links between cutting and grinding and between boiling and crystallisation, land and mill must be co-ordinated, their labour synchronised'.[67] The necessity for initial processing to be undertaken as soon as possible after cutting led to the proliferation of small, seasonally employed production units at the plantation. Although, in the absence of locally negotiated terms, the conditions of royalty

Table 9 Orders for Boulton & Watt engines 1801–25

Year	*Domestic market*	*Caribbean*	*Rest of the world (excl. Ireland)*
1801	47		0
1802	54		1
1803	48	1	3
1804	32	2	4
1805	28	4	1
1806	25	0	6
1807	18	0	2
1808	20	4	0
1809	23	2	0
1810	35	3	3
1811	21	5	3
1812	20	8	1
1813	9	15	1
1814	18	14	2
1815	23	19	1
1816	9	13	9
1817	6	9	4
1818	11	7	9
1819	16	1	2
1820	12	5	2
1821	7	1	2
1822	9	0	10
1823	13	4	9
1824	20	1	6
1825	33	1	4

Source: BRL, Engine Books, letters, Catalogue of Old Engines; J. Tann and M.J. Breckin, 'The International Diffusion of the Watt Engine 1775–1825', *Econ. Hist. Rev.*, 2nd Series, 1978, XXXI(4): 541–64.

payment under Watt's patent would have effectively prevented the seasonal employment of Boulton & Watt engines anywhere in the world until after 1800, the capital and other variable costs of engines would also have been too high at least until *c.* 1815 to encourage intermittent use, save in the Caribbean with its high demand for localized power sources and high animal power costs.

The perceived demand for steam power in the Caribbean was for small, cheap, reliable engines that substituted for inadequate or non-existent water or wind power, or for animal mills, which, in the Caribbean, were particularly costly to run. An 8 hp Boulton & Watt engine with independent framing cost *c.* £998, depending on the number of boilers and duplicate parts required.[68] To this had to be added the cost of transport, erection and engine-house. Fuel costs were low since the engines largely burned trash, the dried cane vegetation and refuse from milling. In contrast to the customer–supplier relationship in the British market, orders for Caribbean sugar-mill engines were placed after minimal technical inquiry or none at all. Thus Boulton & Watt did not incur the costs of becoming involved in protracted negotiations concerning either the application of steam to sugar milling in general or to any roller mill in particular, although they depended on skilled mechanics to erect engines. On the supply side, the application of steam power to sugar cane crushing was probably facilitated by the fact that, by the early nineteenth century, engines were already providing power for rollers in, for example, the ferrous and non-ferrous metals industries.[69] Once the relationship between power input and crushed cane output was known, the prototype was more or less replicated for each successive order. Boulton & Watt supplied 6, 10 and occasionally 12 hp engines: all, save one crank engine supplied in 1804,[70] being independent engines that did not require the more extensive framework needed for crank engines. This degree of standardization was not evidenced by any other market either at home or abroad.[71] Fawcett & Littledale, similarly, supplied a standard product: small 6 or 10 hp engines.

One of the characteristics facilitating the exploitation of the Caribbean market by British engineering firms was the integrated engineering solution to a standard problem or need. In contrast to the British market, where the customer bore a large share of the responsibility to ascertain energy needs, locate local producers of heavy castings and boilers, as well as a millwright and a supplier of the necessary machinery,[72] in overseas markets firms like Fawcett & Littledale and Boulton & Watt frequently supplied an integrated engineering solution. Whereas the former firm made sugar-mills as well as engines, Boulton & Watt did not, although the firm believed that it would have been possible to undercut London founders.[73] Interestingly, it recognized that to diversify into the manufacture of sugar-mills would have been a significant change in manufacturing policy, and not necessarily a desirable one.[74] For a number of Caribbean plantations, Boulton & Watt provided the engine while John Rennie provided the millwork and machinery. Rennie undertook a significant amount of development work on sugar-mill design, receiving communications from millwrights in the Caribbean on the advantages and disadvantages of different numbers of rollers.[75]

> I observe the improvements you propose making in the horizontal sugar mill with three rollers as also that with four rollers. Being still confined to the plantation have not had an opportunity of taking the opinion of any person versant on the subject. But from the trial I have had of the steam engine on this estate with the two roller mills from your works I find no difficulty whatever in pressing the liquor from the canes at one going through. I therefore consider the second pressing as superfluous in a horizontal mill where they have an opportunity of laying the canes all along the roller. But in the vertical mill, where the canes cannot be spread along the roller with the same ease ... the second pressing becomes necessary for the better preservation of the trash or pressed canes.[76]

Table 10 Sugar mills exported by John Rennie

Year	*Number of mills*
1803	1
1804	0
1805	4
1806	0
1807	0
1808	2
1809	6
1810	0
1811	4
1812	6
1813	7
1814	8
1815	14
1816	1
Total	**53**

Source: National Library of Scotland, Rennie Papers.
Note: Some of these were to be worked by cattle.

Nevertheless the three-roller horizontal mill was widely adopted later in the nineteenth century.

Prices for these sugar mills ranged from £490 to £760,[77] although it is not clear whether millwork was included in the latter case. Rennie made his first sugar mill in 1803 and between 1803 and 1816 supplied 53, the majority of which were intended to be worked by steam (Table 10).

The Caribbean was perceived to provide opportunities for ambitious mechanics who assisted plantation owners to implement steam power. William Murdoch, son of William Murdoch Snr, went to the West Indies in 1812.[78] He provided reports on the state of Boulton & Watt engines there and also noted an experiment conducted on Sir Alexander Grant's Jamaican estate to measure engine productivity. Murdoch recommended the use of larger boilers for trash and it became common practice for boilers to be the equivalent of 2 hp more than the engine.[79] Sir Alexander Grant[80] sent a young ex-Boulton & Watt mechanic to Rennie to be trained in millwrighting at his own expense prior to the mechanic emigrating to the West Indies. The climate did not suit all, however. Patrick Blackie[81] reported from Jamaica that he had been asked to go to the south side of the island to erect five or six engines, 'but I am not sure whether the dreadful climate will permit me to avail myself of these offers'. By 1814 there was an iron foundry in Jamaica where parts could be repaired and some replaced.[82]

In both the Caribbean and Indian markets mercantile houses played a significant role in the diffusion of steam power. The East India Company, even after the abolition of its trading monopoly in 1813, continued to play a pivotal role in the diffusion of the Watt engine to India, the company's knowledge of the subcontinent and its mercantile expertise giving it a competitive edge over other merchant houses.[83] In the Caribbean market London, Bristol and to a lesser extent Liverpool mercantile houses such as Thomas Daniel & Son, Inglis, Ellice & Co., Thomas King, Manning Anderton and William & Roland Mitchel acted as intermediaries in the diffusion of steam power, their role as financial guarantors significantly enhancing the attractiveness of the market to suppliers. Third parties, as Mantel and Rosegger[84] have shown, can be standard-setters, evaluating agents, expert decision-makers and advisers as well as early adopters. The British mercantile houses involved in the export of steam engines to the Caribbean mainly acted as evaluating agents and standard-setters in their roles both as selectors of engine suppliers and as exporters, besides acting as financial guarantors to suppliers and, as such, considerably influencing market penetration.

CONCLUSION

Some questions remain. The identity of engine purchasers is problematic. Many engines were ascribed by manufacturers to the merchant third parties who were also responsible for shipping them. It is, thus, not possible to determine whether the earlier adopters of steam engines were the larger plantation owners, whether they included a greater or lesser proportion of absentee owners and whether there was a threshold plantation size only above which was it economic to adopt steam technology. As only the larger plantation owners could afford water-mills it is likely that a similar situation pertained to steam engines. To the extent that larger estates were viable only with a large slave labour force, a connection is made between slavery and steam power. The adoption of steam power technology in the Caribbean in the nineteenth century must be seen in the context of falling plantation profits and the need to drive down costs by optimizing the sugar yield per unit of labour. Slave prices soared in the early nineteenth century in the British colonies and slaves were unobtainable at many locations. (In Cuba, however, slavery continued until full emancipation in 1886.) Steam power was a significant element in the drive for plantation efficiency. The three-roll horizontal mill was genuinely found to achieve higher extraction rates. Towards the end of the nineteenth century, steam power was reliable and cane could be processed, shortly after cutting and before deterioration set in, in seasonally operated mills at the plantation. Processing innovation was matched by agricultural innovations such that labour productivity in Cuba, for example, more than doubled between 1825 and 1850.

The Caribbean was an exceptional overseas market for British steam engines in the early to mid-nineteenth century. Until standardized production of small steam engines forced prices down at the point of supply from *c.* 1815 onwards, the seasonal employment of steam power was commercially viable only in exceptional circumstances. These circumstances prevailed in the Caribbean, where wind and water power were inadequate and the possible alternative, animal power, was exceptionally costly owing to the unusually high number of animals required at each sugar mill. The availability of an integrated engineering solution to sugar milling with standardized small engines and rolls through Fawcett & Littledale or Boulton & Watt with Rennie benefited both plantation customers and suppliers. British mechanics and millwrights residing in the Caribbean and some who were recruited specifically for the purpose of mill and engine erection by larger plantation owners provided the essential practical operative know-how for successful adoption of the new technology. For engine suppliers, the role of mercantile houses as third parties in the transactions eliminated a large element of the risk in exporting engines and machinery. Moreover, Britain's overseas expansion in the Caribbean provided opportunities for the exploitation of protected markets unfettered by competition from other industrial nations or from colonial manufacturers. The colonial status of the market thus both enhanced initial diffusion rates and delayed the emergence of a local engineering industry. For British engineering firms such as Fawcett & Littledale and Boulton & Watt the Caribbean provided a uniquely lucrative prospect. In no other market was there such a confluence of demand and supply factors to facilitate the diffusion of technology on the scale of the Caribbean in the early to mid-nineteenth century.

Notes and References

An earlier version of this paper was given at the 1993 Oxford conference on Technological Change. I am indebted to Prof. P. Cain and Mr P. Hingley for their comments on an earlier draft.

1. A.E. Musson, 'Industrial Motive Power in the United Kingdom, 1800–70', *Econ. Hist. Rev.*, 1976, xxix; C.W. Pursell, *Early Stationary Steam Engines in America* (Washington, DC, 1969).

2. Examples include the mint engines ordered for St Petersburg, Calcutta, Bombay, and the landscape fountain engine ordered by the Nawab Vizir of Oude.

3. A. van Neck, *Les Débuts de la machine à vapeur dans l'industrie belge, 1800–1850* (Brussels, 1979); G. N. von Tunzelmann, *Steam Power and British Industrialisation to 1860* (Oxford, 1978); J. Payen, *Capital et machine à vapeur an XVIIIe siècle* (Paris, 1969); J. Tann and M.J. Breckin, 'The International Diffusion of the Watt Engine 1775–1825', *Econ. Hist. Rev.*, 2nd Series, 1978, XXXI(4): 541–64.

4. J.R. Ward, *Poverty and Progress in the Caribbean* (Oxford, 1985), and *British West Indian Slavery 1750–1834: The Process of Amelioration* (Oxford, 1988); M. Duffy, *Soldiers, Sugar and Seapower* (Oxford, 1987); F.W. Knight, *Slave Society in Cuba during the Nineteenth Century* (Madison, WI, 1970); A.F. Corwin, *Spain and the Abolition of Slavery in Cuba, 1817–1886* (Austin, TX, and London, 1976); G. Cumper, 'Labour Demand and Supply in the Jamaican Sugar Industry 1830–1950', *Social and Economic Studies*, 1954, 2; R. Guerrae and Y. Sanchez, *Sugar and Society in the Caribbean* (Philadelphia, 1964); L.J. Ragatz, *The Fall of the Planter Class in the British Caribbean, 1763–1833* (New York, 1963); D. Watts, *The West Indies: Patterns of Development, Culture and Environmental Change since 1492* (Cambridge, 1987); R.B. Sheridan, *Sugar and Slavery: An Economic History of the British West Indies 1623–1775* (Baltimore, 1973); M.M. Fraginals, 'Africa in Cuba: A Quantitative Analysis of the African Population in the Island of Cuba', in V. Rubin, and A. Tuden (eds), *Comparative Perspectives on Slavery in New World Plantation Societies* (New York, 1977), 187–201; W.L. Mathieson, *British Slavery and Its Abolition* (London, 1926); S.W. Mintz, *Caribbean Transformations* (Baltimore and London, 1974), and *Sweetness and Power* (Harmondsworth, 1985).

5. J. Daniels and C. Daniels, 'The Origins of the Sugar Cane Roller Mill', *Technology and Culture*, 1988, 29: 493–535; N. Deerr, 'The Evolution of the Sugar Cane Mill', *Transactions of the Newcomen Society*, 1940–1 (1943), XXI, 1–9.

6. See also an earlier paper, N. Deerr and A. Brooks, 'The Early Use of Steam Power in the Cane Sugar Industry', *Transactions of the Newcomen Society*, 1940–1 (1943), XXI: 11–21.

7. Based on an estimate of Boulton & Watt's share of the British engine business to 1850, J. Tann, 'Fixed Capital Formation in Steam Power 1750–1850' in C.H. Feinstein and S. Pollard (eds), *Fixed Capital Formation in the Industrial Revolution* (Oxford, 1988), and the scale of Fawcett & Littledale's export business; Liverpool Maritime Record office, Fawcett & Littledale Papers (hereafter F&L papers), Engine books 1813 onwards. The Caribbean was perceived as a potentially lucrative market by other British engineering firms in the first two decades of the nineteenth century but the scale of their operations seems to have been small by comparison with those of Boulton & Watt and Fawcett & Littledale. John Urpeth Rastrick of Bridgnorth Foundry, for example, sold several small engines to West Indian customers through the London merchants Henckel and Du Buisson between 1810 and *c.* 1818, one or more probably going to Martinique. Richard Trevithick also considered the West Indies to be a good market although it is not known how many, if any, engines he exported there; London University Library, Rastrick Papers. I am indebted to Mr Peter Hingley for this information.

8. Watts, *op. cit.* (4), 407.

9. J. Tann and M.J. Breckin, *op. cit.* (3), 541–64; J. Tann and J. Aiken, 'The Diffusion of the Stationary Steam Engine from Britain to India 1790–1830', *Indian Econ. Soc. Hist. Rev.*, 1992, 29(2), 199–214.

10. One hundred and twenty Boulton & Watt engines were supplied to the British cotton industry up to 1825.

11. Watts, *op. cit.* (4), 180.

12. *Ibid.*, p. 112.

13. *Ibid.*, p. 420.

14. *Ibid.*, p. 421.

15. *Ibid.*, p. 413.

16. *Ibid.*, p. 419.

17. *Ibid.*, p. 411.

18. *Ibid.*

19. N. Deerr, *The History of Sugar* (London, 1949), 173–7, 233.

20. J.R. Ward, *British West Indian Slavery, op. cit.* (4), 101.

21. J. Tann, 'Horse Power', in F.M.L. Thompson (ed), *Horse in European Economic History: A Preliminary Canter* (British Agricultural History Society, Reading, 1983), 21–30.

22. Four cattle mills were said to employ 170–200 steers, J. Stewart, *A Plan and Description of Mr John Stewart's Fire Engine Mill* (London, 1776), while an estate in St Domingue had 60 mules in 1788; Deerr, *History of Sugar, op. cit.* (19), 33. See also B. Edwards, *The History, Civil and Commercial, of the British Colonies in the West Indies*, 5th ed., 5 vols (London, 1819); V.M. Satchell, 'Early Use of Steam Power in the Jamaican Sugar Industry, 1768–1810', *Transactions of the Newcomen Society*, 1995–6, 67: 221–31.

23. A plantation in St Domingue (Hispaniola) consisting of 320 acres in 1788, of which 214 acres were in cane, produced 220 tons of Muscavado sugar – at an average yield of 1.14 tons per acre. There were 200 slaves on this estate and 45 mules and 16 oxen for transport, with another 60 mules at the mill. A Jamaican estate was estimated to make a return of 7 per cent on capital investment in 1798, having 250 Negroes valued at £50 each, 80 steers at £10.71 and 60 mules at £20 each. A Cuban estate in 1780 was estimated to return 18 per cent on capital, the capital cost of the whole plantation and buildings, together with 220 slaves, being £35,000. Deerr, *History of Sugar, op. cit.*, (19), 129–233, 335.

24. This was Dugald Clark's patent; see below. For example, Savery claimed that his steam pump could be used for 'occasioning motion to all sorts of millwork', *Miner's Friend*, quoted by Deerr and Brooks, *op. cit.* (6), 11.

25. Patent No. 859, John Stewart, Pamphlet of 1767, cited in Deerr and Brooks, *op. cit.* (6), 13.

26. J. Stewart, *op. cit.* (25).

27. Deerr concluded that 'there is no record of his engine having been built', *History of Sugar, op. cit.* (19), 550.

28. *Ibid.*, p. 551.

29. *Ibid.* That the engine was erected in Jamaica is attested by a French planter, de Cazaud, who referred to it in a paper read to the Royal Society in 1780. Deerr, *The History of Sugar, op. cit.* (19), 551.

30. J. Stewart, *op. cit.* (25). Stewart's engine was made by John Lemon, a London millwright.

31. Birmingham Reference Library (BRL), Boulton & Watt Collection (B&W), W. Pulteney to Boulton & Watt, 1 May 1776 (Box 2). All letters to or from Boulton & Watt quoted hereinafter are from this collection.

32. BRL, Lord Penrhyn to Boulton & Watt, 30 January 1786, 13 February 1787 (Box 34).

33. BRL, George Glenny, Jamaica, to Boulton & Watt, 19 June 1788 (Box 4).

34. BRL, John Dawson, Liverpool, to Boulton & Watt, 9 November 1790 (Box 4).

35. BRL, S. Whitbread, London, to Boulton & Watt, 13 May 1789 (Box 4). Whitbread reported that 'the subject of the Slave Trade was extraordinarily well disrupted in the House last night by Mr Wilberforce'.

36. BRL, Whitbread, London, to Boulton & Watt, 30 March 1790 (Box 4).

37. BRL, Whitbread, London, to Boulton & Watt, 2 July 1790 (Box 4).

38. BRL, Matthew Boulton, Cheltenham, to James Watt, 4 September 1794 (Muirhead Box iv).

39. Knight, *op. cit.* (4), 19. Arango, the Cuban equivalent to France's Colbert, was in public administration for forty years during which time he promoted the acquisition of technological information as well as arguing for a larger measure of free commerce, including slaves. Corwin, *op. cit.* (4), 80.

40. BRL, Matthew Boulton to James Watt, 4 September 1794 (Muirhead Box iv).

41. A reference to the engine 'pirates' Jonathan Hornblower and Bull. J. Tann, 'Mr Hornblower and His Crew: Watt Engine Pirates in the Late 18th Century', *Trans. Newcomen Society*, 1981, 51, 95–109.

42. Deerr, *History of Sugar*, *op. cit.* (19), 15.

43. This was strongly advised by the merchant Fermin de Tastet through whom a number of engines were ordered for Spain. Boulton & Watt had, apparently, asked his advice on credit. De Tastet replied that he never gave credit without a merchant or banker, adding, 'Were we in your place we would go still further – we should not be satisfied with a guarantee abroad, but we should require one in England and if you cannot do foreign business on those terms you will in our opinion do better not to undertake it.' BRL, Fermin de Tastet to Boulton & Watt, 5 September 1794 (Box 36).

44. BRL, Boulton & Watt Engine books; Catalogue of Old Engines. Deerr and Brooks made the total 146 but they included three in Brazil, and one each in Georgia and New Orleans, Deerr and Brooks *op. cit.* (6), 16.

45. Liverpool Maritime Record Office, F&L Papers, Engine books.

46. Watts, *op. cit.* (4), 497.

47. P.J. Cain and A.J. Hopkins, 'The Political Economy of British Expansion Overseas: 1. The Old Colonial System', *Econ. Hist. Rev.*, 1986, XXXIX(4): 501–25, and *British Imperialism, Innovation and Expansion 1688–1914* (London, 1993).

48. J.R. Ward 'The Industrial Revolution and British Imperialism 1750–1850', *Econ. Hist. Rev.*, 1994 XLVII(1): 44–65.

49. D.A. Headrick, *The Tentacles of Progress: Technology Transfer in the Age of Imperialism* (New York and Oxford, 1988).

50. Cain and Hopkins, *British Imperialism*, *op. cit.* (47), 510–12.

51. Ward, 'The Industrial Revolution', *op. cit.* (48), 58.

52. *Ibid.*, p. 59.

53. *Ibid.*, p. 60.

54. *Ibid.*, p. 62.

55. Headrick, *op. cit.* (49), 7.

56. Tann and Breckin, *op. cit.* (3), 546.

57. For example, during the Revolutionary and Napoleonic Wars, J. Tann, 'Marketing Methods in the International Steam Engine Market', *Jn. Econ. Hist.*, 1978, 38: 363–91.

58. Through Manning Anderton, BRL, Catalogue of Old Engines.

59. For Friendship and Saratro Plantations in Essequibo and Belle Plaine Plantation in Demerara, *ibid.*

60. F&L papers, Engine books.

61. Cain and Hopkins, 'The Political Economy', *op. cit.* (47), 522–3.

62. *Ibid.*

63. Only the Spanish patent being granted, Tann, 'Marketing Methods', *op. cit.* (57), 368–71.
64. Ward, *British West Indian Slavery 1750–1834*, *op. cit.* (4), p. 100
65. Sugar consisted of water 69–75 per cent, saccharose 7–20 per cent, reducing sugars 0–2 per cent, fibre 8–16 per cent, ash 0.3–0.8 per cent and organic non-sugar 0.5–1 per cent, Deerr, *Cane Sugar* (London, 1911), 13.
66. Weight loss was rapid. After 24 hours it was 2.19 per cent, after 48 hours 4.03 per cent and after 120 hours 8.57 per cent, *ibid.*, p. 168.
67. Mintz, *op. cit.* (4), 50.
68. BRL, Boulton & Watt to C. Audain c/o Protheroe & Claxton, Bristol, 21 October 1811.
69. J.R. Harris, 'Employment of Steam Power in the Eighteenth Century', *History*, 1967, 52, 133–48.
70. A 12 hp engine for H. van der Henvel, ordered by C. & R. Puller of London for Demerara, B&W Papers, Portf. 352.
71. Engine drawings, titled by the name of the first customer, often have up to six and sometimes more additional customer names and dates in the margin, e.g. Lushington's 1814 engine design also employed for Gilbert Mathison, Robert Allen, France Fletcher, Yates & Co. (twice), Thomas & William King, Kingstons Lambert & Egan, W. & R. Mitchell, Thomas Daniel & Sons, Chas. Lawrence, Campbell Bowden & Co. between 1814 and 1816, BRL, portfolio 872.
72. J. Tann, *The Development of the Factory* (London, 1971), 71–89, 95–105.
73. BRL, Matthew Boulton to James Watt, 4 September 1794 (Miv).
74. *Ibid.*
75. Nat. Lib. Scotland, Rennie Papers, Patrick Blackie, Esher Plantation, to Rennie, 28 June 1814.
76. *Ibid.*
77. For example, a mill ordered by Thomas Daniels in 1813 cost £730. Nat. Lib. Scotland, T. Daniels & Sons to J. Rennie, 3 July 1813.
78. He was said to be able to erect engines and superintend any other sort of machinery. BRL, Boulton & Watt to W. & R. Mitchell, London, 24 February 1812.
79. BRL, Murdoch, Jamaica, to Boulton & Watt, 28 February 1813. It was estimated that a 10 hp engine could produce 7–8 hogsheads of sugar per 24 hours.
80. BRL, Sir Alexander Grant to J. Rennie, 4 March 1813.
81. BRL, P. Blackie, Jamaica, to J. Rennie, 28 June 1814.
82. Ward, *British West Indian Slavery*, *op. cit.* (4), 101.
83. Tann and Aitken, *op. cit.* (9), 206–9.
84. S.J. Mantel and G. Rossegger, 'The role of third parties in the diffusion of innovations: A survey', in R. Rothwell and J. Bessant (eds), *Innovation: Adaptation and Growth* (Amsterdam, 1987), 123–34.

Uneven Mirrors: Towards a History of Engines[1]

MICHAEL FORES

If technology is no more than an arational craft or an application of science, then the history of technology is reduced to a chronological description of artifacts and processes which, before 1700, belong to the prescientific infancy of the human race and, after that time, are derivative of humanity's sole form of intellectual adulthood, science. The claim that science is the only objectively valid form of knowledge leaves technology [here, a process or an activity engaged in by human beings, not a set or records or logs] 'mindless', bereft of its own intellectual method. (Staudenmaier, *Technology's Storytellers*[2])

The hiving-off of cognition and coercion as distinct spheres of [human] activity is at least as important as the emergence of specialization within the field of production. (Gellner, *Plough, Sword and Book*, on the role of the 'clerisy' from the era of 'agrarian society'[3])

INTRODUCTION

Within the first pages of the first volume of 1957 of the *Cambridge Modern History*, a comment can be found from one of those whom readers must assume to have been among the group of self-confessed moderns, about a well known feature of that European age which the 'moderns' have judged to have come before their own in the passing of time. In times widely thought of by the same folk to have been characterized by a superstitious and parochial human outlook on the world, in times we have been told to call 'medieval' in a 'modern era', there were certain changes in the general population in Europe linked with the advent of the bubonic plague and the Black Death. But why should these have been written up in the *History* as comprising a set of 'demographic disasters',[4] given that latter-day scholars, specialists in what is 'cognition' in the epigraph, should have substituted rigour for piety and self-interest? For human beings only, it may be thought, albeit parochially in terms of the many species of the planet, disastrous that there was such a loss of human life. But how many European graphers of any stripe should have perished in, say, the fourteenth century, for such a demise to have been 'disastrous'? A half-dozen demographers, say, out of a score of them in place, of scribes and scholars of population?[5]

There may be those people inclined to argue, when confronted with these queries, that this scholarly author, writing on 'The Renaissance 1493–1520', had simply produced a slip of the pen, to have tagged general population changes as 'graphing' ones. Yet earlier in the same text, about what a separable state of modernity might have comprised, we can read of 'the geographical discoveries'[6] from Europe, events puffed up strongly in the original *History* as those of the Age of Discovery, finding the New World and the 'greatness' of Cristoforo Colombo.[7] So, what kind of grapher was Colombo, who certainly did not take part in what the later *History* has linked in importance with the invention of the printing press as 'the discovery of America' (whatever else he may have done), if only because there was no landmass by that name when he was voyaging away from Europe?[8] Was he a left-handed grapher, prone to using an italic script? Was he a messy map-maker, or a neat one drawing imaginative pictures of fabulous beasts on plain surfaces indicating where unknown territory lay? Or do we confront another slip of the pen from the group of scribes, graphers and loggers, given that the main 'discoveries' of Colombo and his like were not of the form of graphery at all, not articulations on parchment or paper?

The case to be argued follows that of an earlier piece in this series, 'Hamlet without the Prince',[9] suggesting that technical skill has been strangely neglected in histories, if only because the loggers, graphers, scholars and members of the clerisy (*Plough*, cited above) involved have had trouble in getting to grips with it. The use of phrases such as those cited here provides a sign of a certain gentrification in the accounts preferred about human activity in the past, present and future, whereby things occurring away from scholarly locations have been made to seem familiar. 'Biologists' – scholars and so neighbours to the self-styled 'historians of technology' – produce logs about life. Yet those who were discussed as engine-wrights in 'Hamlet', and who have turned up as 'technologists' (apparently log-makers, such as the biologists and sociologists) in both the specialist and general English-language histories, would typically have faced the sack if they had only, at any time in the past, produced logs as their main working output. Self-employed engine-makers would have gone out of business, had they thought of themselves as '-ologists' of any colour.

All the same, such technical specialists have consistently been tagged thus recently, as if they were scribes and scholars, in such books as *Technology's Storytellers*, as quoted in the epigraph, from a study of the articles of the first years of *Technology and Culture (T&C)*, scholarly journal of the Society for the History of Technology (SHOT). In this way, then, aspects of a subject of concern away from the printed page (engines, their making and operation) have gone in danger of being confused and compounded with aspects of a made-up object (indications on paper of this subject). Treatments of such as 'technological progress', 'technological cognition' and an allegedly identifiable 'technological knowledge', as so tagged in writing, sound as if they have been about second-order affairs, not about the primary activities of engine-making. We seem to have more in focus the likes of Colombo's skills as a map-maker than his skills as a voyager, rather as a latter-day discussion of such items as 'economic activity' and the 'economic growth' of human groups has seemed to centre more on the actions of the *nomoi* (stewards or loggers of activity) than on anything else.[10] (How many more economists must be in place in a single country for 'economic growth' to be discerned? A couple of dozen, within *Plough*'s clerisy?).

The author of *Storytellers*, now editor of *T&C*, has made a special effort to 'search out and articulate the language that historians of technology have created to interpret the technological past' (multiple loggings), as invented for developing parts of a 'shared discourse' within a scholarly specialism.[11] But troubles over a chosen verbal usage, over the signifying of something(s) else on to paper, are not all that have to be dealt with, in what turns out to be a 'deep down they are shallow' tradition from the faculty of folk trying to make sense of the actions of others. Take just the identification of 'technological' affairs in the past, whereby specialist and general historians seem to have led each other hand in hand away from grasping what today's practitioners tend to call 'engineering'. The first main theme of *T&C*, for *Storytellers*, has been an Emerging Technology (sounding like the emergence of a butterfly from its chrysalis, without need for human agency), including that of 'inventions'. It is disturbing to learn that the genteel contributors involved have typically shirked trying to grasp the core of making new artefacts, 'the individual creative process' (in 'inventing', here), in favour of studying such peripherals as motives to make things.[12] But what of the character of, say, Emerging Inventions *per se*, given the book's stress on what have been variously named 'design and its ambience' and its 'context', as having been central to 'technology'?[13]

Most people will focus on the 'inventing' of new engines and productive systems, on the making in the main of concrete items, regarding technology-as-hardware – though the book has focused on people doing such as 'creating a new insight'.[14] This seems to have rendered such items as 'technological knowledge', 'cognition' and 'style' – as named differently from 'technical skill' and 'technical design' (no stress on loggings)[15] – to have been concerned with the generation of technology-as-ideas (loggable on to paper), if not with technology-as-engines, as Emerging Inventions might have suggested. With *T&C*'s second major theme, however, Science, Technology, and the Characteristics of Technological Knowledge – hey presto! – the main topic has become technology-as-an-activity, as set out in the epigraph, this being neither of the 'technologies' of its first theme. Then, too, with the third main topic, Technology and Its Cultural Ambience, including 'technology transfer',[16] there is now – horror of horrors, for today's scholars who stress rigour of thought and presentation – yet another meaning which has cropped up of the single English word in the frame, perhaps a mixture of hardware and the know-how needed to operate it. So it is worth pointing out that, with this lack of a tying-down in profane definition of 'technology', its discussants have shown their sense of awe for it (or them), and critics might well class this with 'medieval superstition' about awesome forces.[17]

This signification that 'technology' has been rendered sacred by the scholars, as some have rendered 'science' sacred,[18] can now be treated with

Storytellers on the *T&C* authors on human 'technical skill' (not 'technological skill'), as in their discussion about any links to be found within 'the relationship between science and technology'.[19] Most people will follow the normal first dictionary sense, to allow skill in general to be a personal quality of individuals. *Homo* A has hired *Homo* B partly for what B can do skilfully – from devising histories, to giving legal advice, to making and operating mousetraps and other engines. Skill should thus be contrasted with knowledge, since the latter ought to be articulable on to paper – 'I know X', perhaps to be classed later within 'science' and/or 'X-ology'. Yet the scholarly author of *Storytellers*, as with his supporting of the claim cited in the epigraph about 'arational craft' and the 'applied science' model, has classed 'technical skill' with 'technological knowledge'.[20] So he has made out, against a common perception, such skill to be articulable on paper and similar surfaces. Participants of the *T&C*/SHOT 'shared discourse' have apparently accepted the reality of 'technological knowledge, the necessary mediator between technical design and its historical context', and some have even gone on to 'interpret technological praxis as a *form* of knowledge rather than as an *application* of knowledge'[21] (emphasis in original). The marines, when told this, would fall about laughing.

The case to be expanded below is perhaps best made by focusing on the model of a piece in *History of Science*, 'Homo et Machina'.[22] With active *Homo* having been a forceful, meddling machine-user down the years, and with machinery dynamics having been typically predictable when compared with human dynamics, active *Homo* has been the one of the pair to have specialized in dealing with risk, the unexpected and the unknown. The case, then, cited in the epigraph, about 'arational craft' and an 'applied science' model of 'technology', tends to draw attention away from how real-life human agency has been used in such dynamics, if only because of a false and implicit scholarly concern that admirable human action has been 'rational'. As do, too, *Storytellers*' notion of the apparently effortless 'emergence' of new 'technology' and its statement that 'no technical praxis is *completely* reducible to abstract theory' (emphasis in original)[23] confuse about real events. Here we are in the arena of the posing of misleading 'leading questions', such as with one cited before in 'Hamlet'. About an apparent leadership by Britain in times which 'economic historians' (historians dealing with more than just economics) have tagged as those of 'the industrial revolution' – 'Was it . . . achieved . . . through the effects of "practical tinkerers", or did it have some scientific basis?'[24] (Note the use, again, of the engineers' 'base', for what has been presumed to be necessary for technical achievement.)

Within the real-life enterprises where *Homo* has operated, it has been the machines, productive systems and engines which have been the rational dynamic actors, in the sense that reasons for conduct chosen may be discernible to outsiders. Any dynamic praxis which has been 'reducible' on paper to articulated theory is likely to have been machine-praxis, with scant regular need for human participation. Outside the printed page, as forged by the genteel scholars and loggers, there can have been no separable body of a 'technological knowledge' or an 'engineering science' – should engineers spurn Newton's laws of motion, or those of thermodynamics, because the physicists claim them?[25] Technical skill is not anything readily articulated, or readily understood, by those who do not possess it; accordingly, a clerical classing of technical 'praxis' as a type of 'knowledge' serves to cloud the issue about both. Further, some type of specific 'design', as preparatory work, has been less central to technical activity than a current historians' tradition has made out. Vincenti, contributor to *Technology and Culture* and author of *What Engineers Know and How They Know It*, subtitled *Analytical Studies from Aeronautical History*, may be a likely hero to the clerisy of *T&C*/SHOT, as someone who has admitted never to have designed an aircraft, as someone who has viewed aeronautical engineering from a distance; but he is, as a long-term researcher and engineering professor, an unlikely sage about engine construction and machine use in the raw.[26] Overall, the ill-effects of treating a wispy 'technology' by scholars have been broadly what they might have been predicted to be, from noting the special mode of 'shared discourse' invented for use, discourse that often does not consistently smell of real-life working locations and would not be readily understood by those doing the job in any age.

In the following section, the issue about the centrality of design will be amplified. Forceful *Homo* has tended to 'have designs' on things (neighbouring states, other people's wives ...), in meddling ways not always widely welcomed. Design materials, such as those of 'engineering design', have included centrally sets of plans put down on paper (or the equivalent) about future projects and products, plans typically made away from sites of engine-making and -building.[27] In the overall conditions confronted off the printed page by spirited, adaptable *Homo*, however, plans do not always indicate final outcomes well, as the young of our ingenious species tend to accept readily, but some older heads may have forgotten. That said, within engine-making, some sorts of designerly, synthesizing skills have been needed over the years, as opposed to the mainly analytical, dissecting skills of *Plough*'s clerisy at work. Engine-makers must be able to think in and out of two- and three-dimensional forms, to envisage, in the form of overt designing or not, what future solutions might turn out to be in bulky terms. Here, for success in engine-making, you have to hire those with such skills, or possess them yourself, rather than simply buy books on 'designing things', with designing significantly difficult to plumb. So, it may be easier for outsiders to technical activity to treat its skills as a residual item, after all else has been rationalized by outsiders, rather than as some possible part of *Technology's Storytellers*' notion of technology's 'own intellectual method' (epigraph at the start of the chapter).

Subsequently, I shall try to show how the (human) scholars, clerks and loggers of *Homo* have progressively drifted away from this active animal, at least from those Renaissance times, about which the invention of an ideal human type has been clouded by sightings of 'the discovery of the world and of man',[28] as if this type were not an invention at all. Ideas of an eighteenth-century 'order' and of an 'industrial condition' have provided more chances for the distancing of *Homo* from *Homo*, with the posing of more unanswerable leading questions in the historical literature, such as 'When did technology first become scientific by type?' In such a context, with their insistence on eschewing the verbal styles of the makers and users of engines, the 'historians of technology' have moved too far away from active *Homo* for their own good and that of their readers. So the final section below will make a few suggestions for change for the constructing of more accessible histories, with more stress on engines and the skills of engineering, with less on the dramatics of individual invention and product innovation, with less overtly on the 'contextual history' favoured by SHOT/*T&C*,[29] which has generated too much concern with peripherals.

ARTIFICE AND DESIGNING

Imaginative and forceful *Homo*, as observed in its various antics and actions, as depicted too in the model mentioned above, tends to use words, being abstractions of a sort, in a fairly strictly operated, orderly fashion. Even though much of human discourse may have been jovial or light-hearted and far away from 'information flowing', the use of the names of things (objects, here) has been coined and developed to point securely to the identities of specific things under discussion (subjects). Thus the singular English word 'artifice', as chosen for the subheading here, in a piece of writing under the subtitle 'Towards a History of Engines', has been so chosen because it has been intended to denote broadly what most readers will have assumed, in terms of subject matter outside the written and the printed page. Down the years, and for as long as we have a grasp of *Homo*-not-always-so-*sapiens*, examples of this animal species have deployed their observable artifice to a variety of ends – conquering foes, in sport, in what English-speakers now dignify as 'the arts and sciences', in cooking, in the bizarrely named 'technology'.

One reason for questioning the good sense in choosing the name 'history of technology' for what this affair has seemed to others to be about, concerns acts of naming. More people down those years doubtless have been technical actors than scholars. There have been more of our teeming species engaged in what the author of *Plough, Sword and Book* has rather quaintly called 'the field of production' – just whose factories are in Twenty Acre Field? – than there have been in the clerisy and in *Plough*'s rather grandly entitled 'sphere' of those specializing in 'cognition' – are these folk meant to have inhabited a different planet from the rest of their species?[30] As I tried to point out in 'Hamlet', whose main theme was the non-appearance in published scripts of a proper treatment of the reality of skilful *Homo*, with regard to the example of a 'joiner's' working activity, any overt production and use of such things as logs and designs tends to have decreased with any increase of personal competence. The more competent the individual has become, the less that individual will be inclined to log it on to paper in any form, and/or to make reference consciously to other people's loggings.[31] So why choose the particular name 'technology' to try to deal with all of this striving and effort and its results; when *techne* means 'skill' or 'art', and *logos* (from *lego*, I speak or say) means articulated 'word' or 'law' or 'discourse'; when all the while, virtually anyone who has ever been engaged in what is called 'engine-making' here, will be inclined to claim that articulated words, rules, information and knowledge have rarely lain at the heart of working success?

For pushy *Homo* generally, which can be devious as well as determined, and for groups of this species which has collectively altered its single planet of occupancy significantly, to be said to 'have designs' on something implies, as suggested in the introduction above, the threat of underhand and threatening conduct in the future. Perhaps *Homo* A or Group X has been planning to 'engineer', or 'manufacture', or 'contrive' something to the disadvantage of *Homo* B or Group Y, unfortunate enough to be neighbours of the instigators of such a design or simply to have 'got in their way'. (In Code Homo, as discussed in 'Homo et Machina',[32] use of all these particular verbs tends to imply unfavoured conduct – *Homo* often articulates to praise or dispraise subtly.) An 'artless' designer of such a type might readily be found out in time, for others to frustrate a specific 'design's' fruition; but, as for an 'artful' deviser of artifice, a dodgy character with guile and subtlety, the tale could differ. In these affairs, as in much of what is known nowadays as the 'play' of 'sport', disguise and deception have been of the essence of potential success for clever, scheming and constructive *Homo*, the animal that foolishly knows itself as 'wise'.

The 'shared discourse' of the 'historians of technology' has, if *Technology's Storytellers* is correct, promoted 'technical design' – which readers have to assume, as they have not been warned otherwise, to be what engineers nowadays call 'engineering design'[33] – to have been a rather heroic sort of affair. Most of Vincenti's book *What Engineers Know* ostensibly has been about design and designing – though there has, in the event, been some naming-changing from original *T&C* and other scripts, to suit this particular fad[34] – even though there are few graphic examples of final product-design reproduced in its pages. Layton, another heroic figure for SHOT/*T&C* from outside the workshop, has also been keen to stress the centrality of design – 'design is more fundamental to engineering than is science ... engineering is fundamentally design'[35] – to 'technology', whatever that may be on a Tuesday.[36] Here, now, is what appears as a key passage of writing in *Storytellers*, a volume stressing the importance of 'technical design', but one which has no single graphic example of the genre in it – this is where the member of the *Plough*'s clerisy involved, an apparent specialist in human cognition, has argued that 'the disjunction between knowing and doing' cannot serve the same purpose: 'The tension between abstract and concrete [as produced by engineers] knowledge, between design and its ambience, is the fundamental basis for the interpretation of the history of technology as it has developed' in *T&C*'s authors' work.[37] Note the use here of the engineering metaphor of 'base', for a superstructure which would collapse without its necessary foundation, also the use of 'design' as a shorthand for 'technical activity'.

Consider, now, such a synthesis in the context of the dynamic operation of human artifice in general, as indicated from the 'Homo et Machina' model. Consider too what has also been outlined above about the core actions which *Plough*'s clerisy have been up to, as classifiers and scalpel-wielders of analytical intent. And consider the example of what might have been going on in mousetrap-making, as well as in joinery. (The English word 'artifice', used in the section heading, comes from *artus*, a 'classical' root meaning 'joint', as also being the root for 'arms' of war, Works of Art and 'articulate speech and writing' for *Homo*.) Note as well that, even though all knowledge produced by our species has been of the form of abstraction from its own specified subject matters, in *Storytellers*, 'abstract knowledge' has been used to refer to the type of knowledge produced and developed by people of the type who call themselves 'physicists' and 'chemists' nowadays. So the 'tension' specified ought to be found in the past between (say) 'physics' and 'engineering science' and between 'engineering' and everything else around it.

English words with the prefixes 'con-' (Latin) and 'syn-' (Old Hellenic) from 'classical' roots denoting 'together with', and so denoting the assembly of com-ponent parts, have littered the scene of attempts to put on to paper something about human artifice and cultivation. So too have English words with prefixes from the same two languages, 'trans-' (Latin) and 'meta-' (Hellenic), denoting movements across physical or metaphorical distances. Thus, scheming and devising *Homo* has been the most notable con-structor and trans-ferrer of our planet of a wide range of things. *Homo* has col-luded with *Homo* to found col-leges, polities and gardens, to go to war for con-quest and com-mercial gain, to make and con-test games, to con-ceive the rituals of organized faiths, to make and make use of engines, syn-thetic materials of many types and languages (*T&C*-speak, mathematics, signs for engineering design), in the last of which meta-phor has played a key part. An alleged 'industrial condition' for *Homo* has been flagged as having required new means of trans-port; an alleged 'information age' has depended on the trans-mission of things; scholarship through the ages has required accurate trans-lation of texts. In all of this inventing by *Homo*, in all of this use of artifice (jointing-up), science – even if it has seemed part of the landscape of recent generations of *Homo* – has, like a so-called Work of Art, been as unnatural a thing as any other product of technical expertise. Hence the need, for making 'Hamlet', to imply how this comment from the newer Cambridge *History* – that there may have been the generation of a 'method of discovery that would provide both explanations of nature and a rational basis for technology'[38] – about something linked with the name of Francis Bacon, has been a creature of piety.

In the real world, as opposed to some type of universe invented for the printed page, since *Homo* has been an animal which has con-stantly improved its active praxis – and has so been able to learn new tricks – the possibility is nil that this resourceful creature might have generated, outside its own self-apologies, a discernible 'method' for doing anything actively. Regarding in particular the actions of the mousetrap-maker, unglamorous but useful none the less, if others have been buying his wares, there can have been no helpful certainty that the individual in the frame might be able to 'base' his potentially successful actions on any discernible laws or principles or theories of working practice. Regarding the case of the more glamorous 'scientists' and their forerunners, since no one has ever been able to say definitively what a 'scientific method' has comprised, it is reasonable to conclude that there has been no such thing. Besides this, *pace* the Cambridge *History*, any articulable method (an unbelievable proposition itself) would have to have been other than a 'method of discovery', if only because science and 'natural philosophy' have always been artefacts devised and constructed by human agency. So it is no more believable that 'Newton discovered the laws of motion' linked with his name, than that 'Columbus discovered America', each in times we are told were 'early modern', after an Age of Faith.

Machines have mainly been force-transmitting devices, from the operation of hand tools such as levers and axes to that of sophisticated later machine tools and electronic computers. The animal species, *Homo*, has been a machine user over the same period of the past, stretching up to the present of any time of writing. So, *Homo* has appeared as an especially spirited, wilful and forceful animal species, both as an individual and even more so when acting in human collectives operating with *Machina*, as well as a meddlesome one. This has been the case markedly, as the machines that *Homo* has, for good or for ill, devised, made and put to use have become ever more 'powerful'.[39] Compare the use of such everyday phrases, all in connection with human dynamics, as 'the Force', for the civil police engaged in 'law enforcement'; the Armed Forces; the possession of 'a forceful personality' by some individuals of our species; the collective 'workforce' of companies; Team A 'forcing' Team B into unwanted action in sport; individuals conceived as a spent force; the real-life 'forceful' bullying which goes on at work and in the household, as certain individuals seek local dominance.

Physical forces transmitted within machines have to have been transmitted somewhere other than within the combination of *Homo* and *Machina* themselves. So the main output of machine activity has necessarily impinged on the human surroundings, moulding them, scouring them, 'forming' them anew. There has been nowhere else for all this forceful action to go. In such general manner, *Homo*, with the aid of machines, has sought, often successfully, broadly to impose its designs, plans and dreams, as well as the results of its various campaigns, crusades and enterprises, on to the immediate landscape and the human surroundings. And these affairs of alteration have been classed variously, in recent decades, as 'economic', 'political', 'military', 'social', 'technical' and other types of change. Whatever chosen flag of convenience *Homo* has flown, there has been forceful changing going on galore. Precise plans may have gone awry, but the alteration of the planet has been great.

What, then, in the general context of forceful human artifice and designing, of the following remark from *Technology's Storytellers: Reweaving the Human Fabric*, on 'the relationship between science and technology' which has so obsessed the historians of the latter? – 'Statement 1: Scientific Activity is motivated by curiosity, whereas technology [here, an activity] is motivated by the desire to solve problems.' It has certainly been the case, as the theme of the named book has had it, that the product of the work of the scientists and their forerunners, their 'desired artifact', has had a different form (knowledge) from that of technical people ('engines' in the present sense).[40] But, using the model of active *Homo* developed here, it might be con-cluded that the author of *Storytellers* has been engaged in rather too much imaginative 'reweaving' onto paper about *Homo*'s active past in these respects: to have implied falsely that the scientists and philosophers have not been 'problem-solvers'; to have implied that the engineers and other technical folk have not brought 'curiosity' into their working lives. Similarly, the author of *Plough, Sword and Book: The Structure of Human History* seems to have made too much of a distinction proposed between the active lives of his 'producers', his 'coercers' and the members of a 'specialized clerisy':

> Two specialisms in particular become of paramount importance [in 'agrarian societies']: the emergence of a specialized ruling class, and of a specialized clerisy (specialists in cognition, legitimation, salvation, ritual). ... The pen [of the clerisy] is not mightier than the sword [of the coercers]; but the pen sustained by ritual does impose great constraints on the sword [the swordsmen]. It alone can help the swordsmen decide how to gang up to the greatest advantage.[41]

To try to understand the human past, at least – if not to understand, too, constructed histories (as constructed mainly by *Plough*'s clerisy) – it is preferable to conceive *Homo* primarily in the terms of making and transportation stressed here. Thus the pen has helped to make and produce logs, though it has only helped to do so. The sword has helped to make and produce trade and empires. The saw, the screwdriver and the blueprint have done the same for engine-making – all with the use of human agency, and each being only part of what *Homo* has made. But critically for the present theme, *Homo* has been a forceful animal throughout, with the specialists in cog-nition having made use of co-ercive tactics, in acts of 'ganging up' included, as vigorously as the next person; and *Homo*, maker, has con-structed the pen, the sword, the plough and the book from its own artifice, from this remarkable animal's jointing-together abilities.

Now consider briefly what members of the clerisy of *Plough*'s distinction have been up to generally. No one was known as a 'scientist' before the beginning years of the nineteenth century, and there was no significant group of those whom English-speakers call 'professional scientists' nowadays before the latter half of that century.[42] Whatever various people may have called themselves, however, at any time in the past, and whether they had been paid to do a job or not, there were those engaged in trying to make sense of 'natural' and associated matters and events, in what we have come to accept from the seventeenth century to have been a single world and universe around *Homo*. Though the 'scientists', the present-day scholars and their forerunners are best distinguished mainly from their construction, testing and general manipulation of written texts (accumulations of articulable knowledge), there is something about their actions to be distinguished.

In order to try to understand this activity – so in search of the acts of 'cognition' which *Plough* has had to be a main function of a specialist clerisy – it is worth stressing how such folk have judged a need to ana-lyse (dis-solve or loosen things up) and to dis-sect their chosen subject matters. There has been in view, therefore, the clerisy of the scalpel as well as the clerisy of the pen. Philosophers, scientists, scholars and the like have been the wider society's specialists in active dis-crimination (as in the dis-cernment of separating things apart) within their own special subject areas. Compare and contrast the core synthesis of engine-building in particular and of human production in general. So two other verbal roots besides those cited above ('syn-' and 'con-') are worth citing here, namely 'ana-' (Hellenic) and 'dis-' (Latin), denoting effectively the opposite of the other two – so indicating a con-tinuous scholarly tendency towards splitting up subject matters. That is, the scholarly and clerical splitting up of chosen subject matters (gold/Au, society, engines), in the manner of analysis, scalpel-wielding and loosening things up, in aid thus of future understanding as to be expressed in an object (con-structed logs about gold/Au, sociology, logs about engines).

A dis-eased organism has been an organism assumed, come hell or high water, come heaven or any other condition for *Homo*, to have a contrasted condition from a healthy one (one 'eased' in its ways). Dis-content is the opposite of content, rather as dis-arming a human group is the opposite of arming it, dis-information is the opposite of information, dis-assembling of assembling, dis-order of order. Dis-closure has been something the members of *Plough*'s clerisy have usually claimed to support, if it is linked to science and scholarship. Scholars and clerks alike have to have been constantly mindful of dif-ferences (things borne apart), as well as similarities, in order to hope to dis-tinguish and classify securely, also to dis-course effectively. So too they have placed in their made-up logs identified item A at a dis-tance (standing apart) from item B, as well as themselves at a distance from both named items. They will too, no doubt, have been keen to hear 'dis-covery' stories,[43] as opposed to those of covering up, within a wider effort of scholarship, with its dis-cussions, debates and attempts to loosen up the world prior to cognition.

In this manner, the dis-sections of the scholars, scientists and clerics have been made typically in special subject areas of their interest, as human agency has been deployed to chopping up chosen things in search of understanding about them. They have been the contrasted objects, in forms such as science or natural philosophy or history, which have been the main arenas of constructive human effort, in this general relationship:

subject:object / the past:history / engines:design:human milieu / science

So, clerisy members have chopped up the likes of the biologists' specimens. Then they have devised, designed, invented and made logs of what they have 'found out' thereby. And in this flurry of attempted cognition, the dis-tinctions and dif-ferences propounded by these specialist folk have been as important as the con-structions of their art, including their invention and use of abstractions used in human dis-course,[44] all of which have often been forced by pushy *Homo* on to other people to use, in search of these individuals' own attempts at cognition of parts of precisely the same world as that which the scholars have addressed.

Consider now some aspects of the product of the school of 'the history of technology' introduced above, with forceful *Homo* recognized as having been a producer of artefacts generally, also with what has been articulated here about *Homo*'s general actions within the areas of 'artifice and designing' (the title of the present section). The scalpel-wielders and the pensters of the faculty building have been active classifiers, model-builders, phrase-mongers and makers of the other devices of the arts of analysis. Even to write, think and speak in terms of 'technology' has comprised, by tagging actions which do not primarily make and produce logs together with those of the loggers and graphers, a basic error in discrimination of human activities. This will always have been likely to lead, for instance, to a lay overvaluation of the dependence on logs of all types in what its own practitioners have tended to call 'engineering' for some generations. An explicit focus on 'design and its ambience' among *T&C* writers, for what they have wanted to be central to engine-making, has implied a failure to distinguish between the construction of objects by engine-makers (here, design drawings, specifications, and so on, devised and used to help make final artefacts) and something akin to the making of the subjects of inquiry (here, the engines themselves). Then, too, there has been the oddity referred to above, whereby certain writers have sought, against the run of common experience of *Homo*, to have made human skill a part of knowledge, and to have argued that human technical praxis may have become 'a *form* of knowledge rather than ... an *application* of' the same (*Storytellers*, cited at the beginning of the chapter; emphasis in original).

Among the constructs of those presenting themselves to the world nowadays as 'historians of technology' have been some leading questions, either devised and articulated overtly for trying to make sense of a subject area, or posed covertly for attempted answering in the faculty. These may have been used by the 'historians of technology' to order and arrange their own logs, but they have been forced into the active imaginations of readers of such material nonetheless. Among these has been the question cited above about a foggy 'industrial revolution' in the first sighted instance: 'Was it ... achieved ... through the efforts of "practical tinkerers", or did it have some scientific basis?' (Note again the use of that old warhorse of the latter-day scholars, the identification of a 'base' for a structure raised above it, rather as for Layton design has been 'fundamental' to technical practice as cited above.) Other favoured queries, explicitly or implicitly posed, mainly of times ill-tagged as having been 'industrial',[45] have been these: 'When did science [here, an activity] become rational?' 'When and where did technology [here, an activity] become rational?' 'When and where was scientific method developed?' 'When did technology [an activity] become scientific by type of operation?'[46] 'When did technology [an activity] start to become based on science [an activity]?' 'When did technology [an activity] start to become based on science [an assembly of knowledge]?' 'When did technology [an activity] start to become based on engineering science?'

The operation of the 'Homo et Machina' model, however, suggests that two things are certain about the activities of the scientists, the engineers, their forerunners and everyone else, in all of the historians' various 'ages', 'epochs' and the like. Any well-considered 'basis' for working will have lain elsewhere than in human 'rationality' of mind or spirit; and there has been no distinctive, identifiable 'method' brought about for *Homo* to do anything at all in the dynamic ways discussed. Some of the favourite phrase-making, therefore, of the ill-named 'historians of technology' has not been about real-world events at all; it has not been about anything outside the printed page and the faculty walls. This is where there has been no set of known criteria to contrast *Technology's Storytellers*' 'abstract and concrete knowledge', where an 'engineering science' has been no more securely identifiable than a 'physics' or 'chemistry'. This is where, as well, any identifiable 'tension ... between design and its ambience' (*Storytellers*, cited above) has been only one of scores of such affairs in real-life operation. For the mousetrap-maker, for instance, affairs which others will choose to call those of 'tension' and 'pressure' will have been of the essence of working normality, as customers have asked for what cannot be made for the cost, as suppliers have asked for 'ruinous' prices, as the 'workforce' has gone to war with 'management', as 'development engineers' have dreamed up new materials which do not do the jobs promised, and so on, and so on.

DISTANCE, INDUSTRY AND 'THE MODERN WORLD'[47]

'Modern society is industrial society', the *Britannica* has had it about the 'modernization' of human groups; and the author of *Theories of Industrial Society* has stressed how these theories have been 'centrally concerned with the novelty and power of science and the social consequences of its industrial application'.[48] Such a link proposed between something(s) non-specific called 'science', 'industry' and an apparently separate 'modern' condition may seem to have been confirmed in *Plough, Sword and Book*. This has been, with its claim articulated about the condition of Industria, conceived as if it has been a separate 'scientific-industrial world' characterized by 'scientific/industrial production', conceived in effect in this particular book as the Age of the Book. So it has been proposed as a distinct condition, for the human occupancy of our single planet, from which magic and the sacred have been banished, as something 'based ... on the generic ... discovery [cf. "scientific discovery"] that successful systematic investigation of Nature, and the application of the findings [cf. "scientific findings"] for the purpose of increased output, are feasible, and, once initiated, not too difficult'.[49] Note another 'discovery' story, also another use of the engineering metaphor 'base', for a superstructure which would collapse without its foundations – here, about affairs widely thought nowadays to have followed 'the industrial revolution' of the invention of 'economic historians'.[50]

Accordingly, the 'applied science' idea for understanding technical activity, of which *T&C* authors have come to disapprove for what they have opted to call 'technology',[51] has come strongly into the common human imagination. An example is the way in which libraries in the English-speaking lands have opted to shelve their volumes confidently under such main headings as 200 Religion, 500 Science, 600 Applied Science (including Technology), 700 The Arts. Another is the way in which the state in Britain has chosen to house, since 1905, its main national collection of engineering products in an institution puffed as the Science Museum, whose recent slick advertising logo has incorporated two fallacies: that producing science is mainly about knowing mathematics, and that engineering is mainly about using science. In yet another example, the authors of *The Technology of Man* discuss, within their treatment of the industrial revolution, its inventor's making of the crucial Watt steam engine as his 'discovery' – and so, by implication, necessarily a creature of the (various) thing(s) called 'science'.[52]

We have here a wonderful, organized mechanism in view for *Homo*'s continuing misgrasp of *Homo* from paper-borne accounts, whereby more and more portentous nonsense can slip by under the genteel heading 'scholarship'. When the available facts from the past have been studied in detail, for instance, it has been clear for decades that 'the industrial condition' of nations has not been one in which factory-working has been typical for active *Homo*; neither has it been one in which the clerical 'economists'' 'national product' has stemmed mainly from manufacturing.[53] It is preferable, however, to return at this point from the murky and indistinct affair called 'industry' nowadays, with its opaque walls that have prevented the casual observer from knowing the detail of what has been going on therein, in favour of dealing first with some aspects of what was introduced above as a type of proposed Renaissance-era 'modernity'. The author of *Plough* has, as already noted, discussed the membership of a 'specialized clerisy (specialists in cognition, legitimation, salvation, ritual)'. So the main query of the present section can be phrased thus – how did the clerks in unholy orders of the West come progressively, in times they have insisted on calling 'modern', to drift away from understanding skilful, active *Homo*? And it is worth starting from the time of that famous 'discovery of man' cited in the introduction above.

Cristoforo Colombo, a seafarer rather than a geographer, failed to discover America, for reasons set out above, in the last years of the fifteenth century. The character of an allegedly dis-tinct 'modern' era, also often shown by the incidence of the printing press – for Koestler in *The Act of Creation*, an example of remarkable and miraculous 'scientific discovery' – has become rather muddied of late, as pointed out at the start of the newer Cambridge *History*.[54] Perhaps, indeed, no major new era for *Homo* occurred or started in the years around AD 1500 in parts of Western Europe, as implied by the author of *Plough*, with his thesis of two major 'transitions' in our species's past, the shifts from hunting and gathering to Agraria, then later that to Industria.[55] There were, in the human past, 'the two very great transitions – the neolithic and industrial revolutions ...'[56] Yet, all the same, and whatever the historians' fashion of the day may have been, the idea of the European Renaissance as a great watershed in the general human culture has been with our species for generations; and it has percolated into formal teaching in the West, as linked with the name of Leonardo da Vinci, as linked with the proposition cited above of 'the discovery of the world and of man' and the idea of Humanism. One series of textbooks, for instance, designed to cover much the same ground as the Cambridge *Modern History*, though published under the series title *The Rise of Modern Europe*, has even addressed 'The world of Humanism 1453–1517',[57] about what was really only a minority cult in a very few urban spots within the years specified. What relevance might all this cultural construction have today to the theme of the present piece of writing?

A part of the proposition of an apparently secular Art of those 'modern' times given by the European Renaissance has involved awarding a special insight to people named as Artists, folk to be celebrated, if the makers of Works of Art have been judged competent enough, as 'geniuses' by bourgeois Western Art-lovers. As part of this particular cult, we have been told very often how Art mirrors Nature, or Art reflects Life – with the human type called the Artist having a unique, interpretative role to play. Note the sign of awe involved, to confirm the suitability of the designation 'Art-lover', with the use of the present tense for past events – 'Leonardo is telling us this ... The Artist is saying that ... Shakespeare knows what it is to suffer ...'. The Artist lives on![58]

Now take *Ideal City*, or perhaps it should be given in English as *Ideal Town*, a painting by one called an Italian master of Renaissance times by cult members, as used to illustrate the texts of both Plumb's book on the Renaissance and Kenneth Clark's *Civilisation*, on culture and Art in the West.[59] By the rules sketched out here, the ideas of the newly wrought Humanism of so-called Renaissance times of 'rebirth' of a 'lost culture', as linked partly with an attempt to return to the prized culture of a 'classical' past, ought to be displayed in paintings such as this. And they surely are so. In the painting named, there is no single human being to be seen. 'Urbs without *Homo*. Discuss.' This is rather as in Renaissance-era paintings of the countryside, conceived and created to appeal to selected members of the richer groups of townsfolk; there have been rarely to be found in sight ragged peasants, or the suspicion of the smells of rural living, for townsfolk, say, relaxing on an idyllic and genteel picnic out there *in rure*.[60]

So the truth will out, as exhibited about an influential, formalized Humanism, and as revealed by Art. Eureka! We have a genuine discovery in view! Here has been, from its outset, an outlook on the world, a Humanism, proposed from the study, with little direct concern for *Homo* in general. 'Humanism without *Homo*. Discuss.' The famous Renaissance-era 'discovery of man', in the event, comprised the invention (not the finding or discovery of something already there) of a leisured urban type, for whose newly fashioned lifestyle those who have worked, prayed and fought were set apart. In particular, for the present theme, New Man should be located, where its real-life examples habitually paraded in their finery, distinctly bourgeois and male of the species, out on the piazzas of a few well-known Italian towns and cities. The 'discovered' type called Man was out there – discussing Art and Life in a 'civilized' way perhaps – far distant both emotionally and physically from the women and children left behind at home, far distant from the countryfolk and artisans who provided, in such times, both the majority of the species *Homo* and the physical means to support a leisurely lifestyle of Humanism for the few.[61]

Science, often associated with a human 'objectivity' of outlook on the world and the decline of an Age of Faith, has provided, for the intellectuals of the West, the other great fad besides Humanism of the last five centuries or so. There should have been, at least in seventeenth-century Europe, 'the depersonalization of natural knowledge', as well as the slackly named 'mechanization of nature' (in fact, a proposed mechanization of the logs set down about the human surroundings).[62] So a case could be made at this point for treating ideas of what 'historians of science' have chosen latterly to call 'the scientific revolution', centring on that century in parts of Europe; but it is preferable to select here a synthesis about the legacy of Newton and the others, by making focus on (an) idea(s) of (an) influential eighteenth-century world-view(s) remaining in force today.

The author of *Plough*, under the subheading 'The Terminus', has discerned what he has called 'the new world', generated in that time of the human past comprising his idea of 'a unified orderly world', in which 'nothing is *necessarily* connected with anything else', one in which 'there ... is and can be no room either for magic or for the sacred' (emphasis in original). In this,

> if reference to reality, the collation and ordering and prediction of facts, is its sole and ruling principle, then evidence is king ... This, in very rough and general outline, is the world we inhabit, as codified by our philosophies, or some of them.[63]

That is, we have to presume, from what has been written, about 'the terminus' in the development of human culture. Yet, potentially in stark contrast to this particular assumption, a few pages on, in the same chapter and now under the subheading 'The Overall Plot' (of a book called *Plough, Sword and Book*),[64] there is this statement:

> The most important feature [of 'the scientific-industrial world'] is this: the environment within which man now [1988] lives is unified into a single continuous Nature, assumed to be law-bound and homogeneous, devoid of any privileged ('sacred') elements, open to sustained and never closed exploration, tending towards ever more general and conceptually centralized explanation.[65]

Compare the use of an everyday phrase 'the nature of' to denote what has been assumed from the outset to have been the uniform, unchanging character of the thing named; and note that, for the human author of *Plough*, his own species should, for all its known 'evolution' in the past, be included within that same orderly Nature, relatively uninfluenced by the passing of time.

Well, perhaps so, about all these orderly surroundings for *Homo*. But what if *Plough*'s 'facts', which have been meant, according to this book's Core Proposition One, to 'rule' and lord it over us all in this particular 'scientific-industrial' culture, were to deny, at any future date – and about any specific set of things around us, such as the human dynamics discussed in the present piece of writing – the prior assumption of the same book's Core Proposition Two, that the singular world is orderly and explicable? *Pace* the author of *Plough, Sword and Book*, it seems that he has announced in his own book what can be dis-tinguished, in effect, as two rival 'sole and ruling principle[s]' in operation in a single human culture of Industria, without advising his readers of this potential dis-traction for understanding it. In particular, it has involved a reckless assumption by *Homo*, to have assumed wilful and active 'man' to have been a proper part of an orderly, unshifting Nature, knowable in the manner of the 'centralized explanations' of a post-Newtonian, Enlightenment ideal which has been so very attractive to clerisy members.

The argument of *Plough* has focused as well on what has been discerned to have been a major cultural 'tension' between local and centralized world-views and cultural types among the groups of *Homo*.[66] Once, according to the book's thesis of the working-out of this cultural tension, there were small groups of *Homo* on the planet, groups with specific religious faiths and whose members owed allegiance to dynastic rulers and their like. A degree of centralization came about, by this thesis, with the advent of 'salvational' religions, such as Christianity, with their stress on universal applicability for the species *Homo*. Besides this, by the same thesis, there has been a key role for the clerks in all of the change:

> The central device for presenting doctrine as trans-social, trans-ethnic, as endowed with extraneous legitimation, is to have it located in writing, *and* to have it endowed with a meta-theory, a theology, which

> circumscribes, defines and limits the sources of authority.[67] (emphasis in original).

Here, then, is the stuff for an explanation of the apparent contradiction noted above. The clerks in unholy orders have, always and not surprisingly, been keen to stress the key role of writing in the development of human culture in general. But, in order that they might try to do this, they have had to invent something(s) of the type of *Plough*'s 'meta-theory, a theology'. In particular here, there have been, first, the assumption of order in Nature and associated things and, second, the main fabric of a viewing of the past, such as with ideas about how 'modernity' came about via the 'neolithic', 'scientific' and 'industrial' revolutions.

Concerning *Plough*'s idea of a general 'switch from concept-implementing cultures to instrumental-rational ones':

> The most important difference between the world we have left behind and the world we now inhabit ... is the status of concepts in either. In the first of these worlds, concepts were sacralized ... [as] markers of thought and conduct, and the markers were hallowed by ritual. In the new world, concepts, like men, were levelled out; none were [*sic*] hallowed, all were treated instrumentally, all were expendable and replaceable.[68]

Not so, however. For there is much evidence to hand in our so-called 'scientific-industrial world' of precisely the sacralization of concepts, which the author of *Plough* has said (when stressing 'order' around us, at least) should have been left behind in the human past: the way in which the notion of an 'industrial era' has been protected from the profanations of critical inquiry, the way in which fuzzy notions of 'science' and 'technology' have been have gone relatively unchallenged by pious scribes and clerks.

Regarding what has been called above the codification of a 'different world' which Westerners inhabit today, it is best to assume that the eighteenth-century settlement has not occurred in either of its forms cited above from the book called *Plough, Sword and Book*. Certainly, Western intellectuals have often expressed their self-satisfied belief in order to be found in the future, with the human explanations of some type of 'natural world',[69] and they have also asserted regularly that factual evidence ought to rule in 'modern'-day discourse. Yet close observation suggests that something akin to the 'localism' of 'world-viewing' by members of our species, as set out in *Plough*'s idea of cultural tension with a centralized world-view, has remained the ruling outlook for most of our species most of the time, concerned as they have been with the particulars of place and time – that is, the outlook of people, from whom the clerisy have drifted away in general outlook on the human surroundings, even in what has often been called 'the Western world'. In particular, if it is assumed that *Plough*'s general cultural tension has not worked out as its author has said, that a separable condition (or 'world') of 'modernity' has not arrived to our single planet, and that most people have retained the localistic outlook outlined here, it is feasible to accept this: they have been the heterogeneous, dis-orderly features of skilful, 'gifted' and 'able' *Homo* which have been most instrumental in forging the human culture and surroundings, whose story of development over the years has not ended up with Max Weber's conditions of 'instrumental rationality' for *Homo*.[70]

Readers of these paragraphs must be aware of their own special and 'practical' skills and competences in their own special working arenas. So they might wish to apply the human 'powers of imagination' to the matters of engine-building and science construction, to which the questions quoted near the end of the previous section have been directed. Technical people in particular, and active *Homo* in general, have tended to be successful in sorting out what latter-day scholars have formalized as 'underspecified problems', by becoming obsessed with particulars and specifics, not with the laws and principles and other generalities which the scholars have been so keen on. What members of the Western clerisy have wanted to class in their books as 'basic' for human activity, such as the 'rules' of this or that, or the scientific knowledge of this or that, have normally turned up, in the cold light of the morning, as relatively unimportant items in real-life working conditions of what was called Homocraft before. *Pace* the author of *Plough*, there has been no discernible latter-day 'discovery' that 'systematic investigation of Nature, and the application of the findings' have become crucial in all that 'industrial' activity we read so much about in genteel books and learned articles. So, regarding the leading questions cited above, since not even the scientists and the natural philosophers have been able to develop a 'scientific basis' for their own dynamic operations – in the sense that there may have been some set of written articulations for them to turn to for advice on 'best practice' – it is reasonable to suppose that these questions provide a mare's nest on paper of misleading cognition, about what may have been going on in the factories and workshops in what turns out to have been a grossly mis-tagged 'industrial condition' of nations.[71] The author of *Technology's Storytellers* has simply been wrong to have located major technical 'tensions' where he has done so. In terms of real-world actions, the whole of the first epigraph at the start of the present chapter, as taken from an aptly entitled *Storytellers*, forms part of the same mare's nest. The grandest of the 'findings' of the 'historians of technology' have been among the trivia of our era; they have been the equivalent, when viewed from outside the faculty walls, of horses flying and unicorns at bay.

IT'S ABOUT TIME

Citing the subtitle of the book called *Plough, Sword and Book*, whose author has sought to clothe up in effect his Industria as the Age of the Book, there may well have been a discernible 'structure of human history' to be found on library shelves and in academic debates. But what has gone before here indicates that, at crucial locations, this has not helped to inform well on any discernible and structural aspects of a human past, a past which has been more heterogeneous than the scholars have liked to believe. That is, with the past identified as a different thing from histories, rather as an Artist's subject has always been different from that person's object, as gold/Au has been different from constructed logs about gold/Au, as an engine has been different from an 'engineering design'. The theme therefore of the third section here, concerning *Homo*'s progressive distancing from *Homo*, is broadly consistent with that of an earlier piece in *History of Science* entitled 'Newton on a Horse', in which it was suggested that the newer Cambridge *History* has produced, in effect, a covert eschatology about the various things called 'science' and 'technology'.[72] We have had an unbelievable and Whiggish tale of general progress through the past to a present which has not yet existed. It has been this theme about the human past that has underlain key parts of the construction of Western histories, being one that has been rendered sacred and thus inviolate, closed to critical inquiry. Agriculture and the Watt engine, inventions by clever, resourceful *Homo*, have been cited in *The Technology of Man* as 'discoveries' in their first instances, because these have been presented as part of a sacred tale of how *Homo* has got to an imaginary, relatively free, 'present' condition for the species.[73] Newton's invented calculus has been presented as a 'discovery' by his twentieth-century, New World biographer for the Cambridge press,[74] and no one seems to have noticed this absurdity of classification, because Newton and mathematics have been awarded, by members of the Western clerisy, awesome roles in the paper-borne dramatics about the development of the human culture.

In order to understand active *Homo* well, it is better to tear up and throw away many familiar and beloved abstractions of the scholars' artifice and design. Recycling the author of *Plough*'s engineering metaphor, the Renaissance and Humanism provide unstable columns with which to help support an informative, book-trans-mitted framework for grasping a 'modern'-era past. The scientific and industrial revolutions, together with a shift towards some type of 'scientific technology' of the invention of the bizarrely named 'historians of technology', are poor struts or ties to keep together anything which might inform on the matters raised by those who have invented these concepts. Worries expressed in *Technology's Storytellers*, as cited at the head of the present piece of writing, about the possible lack of a specific 'intellectual method' for technology-as-an-activity indicate how far the clerks of SHOT/*T&C* seem to have strayed from the realities of *Homo*'s working surroundings.[75] When did a representative Dr Robert last visit a workshop for more than a fleeting visit?[76]

With the devising and the framing of the present remarks, we are back to what was articulated in this series at the end of a piece called 'Francis Bacon and the Myth of Industrial Science', where I tried, in a conclusion expressly on Practice by *Homo*, to list some of the conceptual flotsam of our age.[77] In the event, the case of Bacon's prophetic legacy provides a nice example of the sort of 'tribal' distortion of which he wrote so graphically concerning 'the idols of the tribe', part of whose expression has provided the title for the present paper.[78] Uneven mirrors have been well to the fore in suggesting, say, that Bacon was the prophet of 'industrial science'; whereas the real man warned against those cursed with Atalanta's folly, stooping to pick up the golden apple, 'interrupting their speed and giving up the victory'.[79] The real Bacon warned against precisely such types of bookish nonsense as that of a 'union of science and technology', a proposition seeming true in the twentieth century, if only because of its familiarity in print:

> The human understanding, when any proposition has been laid down ... forces everything else to add fresh support and confirmation. ... With regard to authority, it is the greatest weakness to attribute infinite credit to particular authors, and to refuse his own prerogative to time.... For truth is rightly named the daughter of time, not of authority.'[80]

Concerning 'the instances of the road, ... men are wonderfully indolent ... consulting nature [and other things] in a desultory manner, and at periodic intervals, when bodies have been regularly finished and completed, and not during her work'.[81] Going off on a metaphoric road with Bacon and Time, away from the study to where worldly circumstances may be examined in detail, a well-constituted history of engines ought to be focused from the outset on the engines themselves. These have been useful, interesting and changing devices playing, for good and for ill, important parts in forging the general human milieu and our species's culture. They have been fashioned skilfully by *Homo* for a variety of reasons linked in with what scholars have recently chosen to know as 'external' and 'contextual' purposes for their making; some of such engines remain for close scrutiny; and with such direct attention on them, it is not necessary to start off with inherently speculative issues such as any link discernible between particular engines and the 'methods' deployed in their construction and operation, between the making of engines and any knowledge utilized by the makers.

Finally, if readers of this piece are worried that such an approach to dealing with the past will turn out to be disorderly celebration of the human use of its technical skills, the tone of the piece ought to show that it is possible, using such an approach, to be sceptical about human intents at 'designing' and fashioning artefacts and products. As to possible ways of ordering and modelling a complex tale, it is relatively easy to use the passing of time as one parameter. As already suggested, engines have been made in a cultural context. Leonardo da Vinci was a non-inventor of the submarine, because he could not make one and because there seems to have been little demand for such a device at the time. Machines form a group of what have been called engines here, products of technical ingenuity; and the former can be studied fairly readily in terms of their physical characteristics as force-transmitters in a single world much influenced by human habitation.

Notes and References

1. 'The idols of the tribe are inherent in human nature and the very tribe or race of man; for man's sense is falsely asserted to be the standard of things; of the contrary, all the perceptions both of the senses and the mind bear reference to man and not to the universe, and the human mind resembles those uneven mirrors which impart their own properties to different objects, from which rays are emitted and distort and disfigure them.' Translated from F. Bacon, *Novum Organum* (London, 1620), Bk 1, Aphorism 41.

2. J.M. Staudenmaier, *Technology's Storytellers: Reweaving the Human Fabric* (Cambridge, MA, paperback ed., 1989), 102.

3. E. Gellner, *Plough, Sword and Book: The Structure of Human History* (London, 1988), 17, 21.

4. D. Hay, 'Introduction', in *The New Cambridge Modern History* (Cambridge, 1957), Vol. 1, 1–19, p. 3.

5. Cf. Gellner, *op. cit.* (3), 70, on a 'demographic explosion' of the general population, which helped to spawn 'a class of ritual specialists' among other human groups. The author was referring to an idea of a general population change.

6. *History, op. cit.* (4), 2.

7. *The Cambridge Modern History* (Cambridge, 1902), Vol. 1, 7–66, p. 21. Chapters 1 and 2 have the titles cited here. Volume 1 in both cases – *The Cambridge Modern History* and *The New Cambridge Modern History* – deals with the European Renaissance expressly.

8. *History, op. cit.* (4), 4. The earlier *History* tells of how 'an obscure printer' and map-maker named a continent after Amerigo Vespucci (the Latin form of whose name is Americus) partly as a joke in 1507, after Colombo's death, *op cit.* (7), 35.

9. M. Fores, 'Hamlet without the Prince', *History of Technology*, 1994, 16: 161–84.

10. Cf., for instance, Gellner, *op. cit.* (3), 21, on a separate 'sphere' entitled 'the economic', which is concerned with 'production', rather than what it seems to have been, stewardship, from the label selected for use.

11. Staudenmaier, *op. cit.* (2), 2, 34. On the latter page, it has been suggested that *Technology and Culture* authors have produced such a discourse.

12. *Ibid.*, 41.

13. E.g., *ibid.*, 105, 120. Cf. p. 103, on 'the tension between technical [not technological] design and its ambience, which defines the nature of technology itself'.

14. *Ibid.*, 41.

15. E.g. *ibid.*, 103, 120, 200, 114, 167.

16. *Ibid.*, titles of chs 3 and 4. The main themes of *T&C* have been described, at p. 37, as 'emerging technology, characteristics of technological knowledge, and technology and culture'.

17. There are, too, 'historians of technology' who have stressed, sometimes at least, their main subject of concern to take the form of technology-as-knowledge. Cf. E. Layton, 'Technology as Knowledge', *Technology and Culture*, 1974, 15: 31–41, and the title of the first chapter, 'Introduction: Engineering as Knowledge', of W.G. Vincenti, *What Engineers Know and How They Know It* (Baltimore, 1990), 3–15, where, bizarrely, 'engineering' has been defined otherwise, at p. 13, as 'a form of human activity'. This author's 'technology' has also flip-flopped between taking the form of knowledge at p. 5 and that of an activity engaged in by *Homo* at p. 14.

18. Cf. my paper 'Another Dark Labyrinth', unpublished MS, in criticism of the 'historians of science', on sloppy naming as an indication of things rendered sacred, also 'Hamlet', *op. cit.* (9), 166–71, on Science and Horses and Engines and Things, where science has been discussed as a non-valuable 'public good'.

19. Staudenmaier, *op. cit.* (2), ch. 3, esp. p. 83. Cf. Gellner, *op. cit.* (3), 17–18.

20. Staudenmaier, *op. cit.* (2), 114–20, within a section entitled 'Four Characteristics of Technological Knowledge' in ch. 3.

21. *Ibid.*, 102, 120.

22. M. Fores, 'Homo et Machina: Of Divided Labour, Revolutions and a History of Marvels', *History of Science*, 1994, 32: 63–87.

23. Staudenmaier, *op. cit.* (2), 115, in the discussion of 'technical skill', the cited statement as being 'inherent in the design-ambient tension of technological cognition'.

24. A.E. Musson and E. Robinson, *Science and Technology in the Industrial Revolution* (Manchester, 1969), 65, as discussed in Fores, *op. cit.* (9), 164.

25. 'The engineering sciences consist primarily of statements [made by human agency] about human artifacts.' E.T. Layton, 'Science as a Form of Action: The Role of the Engineering Sciences', *Technology and Culture*, 1988, 29: 82–97, p. 97. But this could as well be said of parts of 'chemistry' and all of what has been filed as 'current electricity' under the main heading 'physics'. Cf. M. Fores, 'Transformations and the Myth of "Engineering Science": Magic in a White Coat', *Technology and Culture*, 1988, 29: 62–81, esp. p. 73, as composed, in part, in criticism of a theme expressed by Layton in that journal.

26. 'I have never attempted to design an airplane in my entire career as a research engineer (though I participated in planning and designing large aeronautical research facilities).' Vincenti, *op. cit.* (17), 7. However, chs 2–7 (of 8, in total) of his book are all expressly about design.

27. '"Design," of course, denotes both the content of a set of plans ... and the process by which those plans are produced. In the latter meaning, it typically involves tentative layout (or layouts) of the arrangements and dimensions of the artifice ...' *Ibid.*, 7.

28. Cf. *History, op. cit.* (4), 2, citing Jakob Burckhardt's thesis of 1860 about the Renaissance culture in Italy.

29. Staudenmaier, *op. cit.* (2), xxiii–xxv, 11–12.

30. See the wording of the quotation in the epigraph.

31. Fores, 'Hamlet', *op. cit.* (9), 179–80.

32. Fores, 'Homo et Machina', *op. cit.* (22), 73.

33. E.g. Vincenti, *op. cit.* (17), 7, 12.

34. Chs 3–6 were first published without the word 'design' in their titles; but they have now turned up, rather as if they had been composed specifically about 'design', under the titles 'Establishment of Design Requirements', 'A Theoretical Tool for Design', 'Data for Design', 'Design and Production'.

35. Layton, *op. cit.* (25), 89, 90.

36. Cf. Staudenmaier, *op. cit.* (2), 89, 93–4. On Layton's fuzzy grasp of 'technology' and 'engineering science', see also Fores, *op. cit.* (25), to which Layton, *op. cit.* (25) was a response.

37. Staudenmaier, *op. cit.* (2), 120, in the book's section 'Technical Skill' in ch. 3, expressly on science, technology, and the consequences of technological knowledge.

38. A.C. Crombie and M.A. Hoskin, 'The Scientific Movement and Its Influence 1610–50', in *The New Cambridge Modern History* (Cambridge, 1970), Vol. 4, 132–68, p. 133, criticized in Fores, 'Hamlet', *op. cit.* (9), 169. Note the anachronism of the *History*'s chapter title, since people did not normally think in terms of 'science' in the years specified, but more commonly of 'natural philosophy'.

39. Cf. Fores, 'Hamlet', *op. cit.* (9), on 'the most forceful, wilful and meddling organism of them all', esp. pp. 161, 172–7.

40. Staudenmaier, *op. cit.* (2), 86–95, esp. pp. 86–7.

41. Gellner, *op. cit.* (3), 17–18, 64–66, 99. Quotations from pp. 17, 99.

42. Cf. Fores, 'Hamlet', *op. cit.* (9), 177–82, also on 'the faith of 1874' of the year of the establishment of the Cavendish Laboratory in Cambridge, England, used as an apt symbol of the increasing separation of 'science and technology' of the time.

43. Cf. the discussion of false tales of 'discovery' in my piece 'Constructed Science and the Seventeenth-Century "Revolution"', *History of Science*, 1984, 22: 217–44, pp. 221–5, as taken up again in 'Another Dark Labyrinth', *op. cit.* (18), in criticism of an organized 'history of science'.

44. Cf. A.N. Whitehead, *Science and the Modern World* (Cambridge, 1926), ch. 1.

45. For a critique of the idea of an identifiable 'industrial condition', see my pieces 'The Myth of a British Industrial Revolution', *History*, 1981, 66: 181–98 and 'A Cambridge View of Modernity', *J. Hist. Sociology*, 1992, 5: 209–33, pp. 224–9.

46. Cf. E. Layton, 'Scientific Technology, 1845–1900', *Technology and Culture*, 1979, 20: 64–89.

47. Cf. Whitehead's given title, *Science and the Modern World, op. cit.* (44), about something never really a 'world' apart, in the sense of different planets, but rather to do with the suppositions of the age-mongering historians and other scribes.

48. K. Kumar, 'Modernization and Industrialization', in *Encyclopaedia Britannica* (Chicago, 1992), Vol. 24: 280–91, p. 280; R.J. Badham, *Theories of Industrial Society* (New York, 1986), 127.

49. Gellner, *op. cit.* (3), 17–20, 64–6, 68, quotation from pp. 17–18. Cf. the notion of the

'union of science and technology', as discussed in D. Birdsall and C.M. Cipolla, *The Technology of Man* (London, 1980), 138.

50. E.g. C. Beard, *The Industrial Revolution* (London, 1901); P. Mathias, *The First Industrial Nation* (London, 1983).

51. E.g. Staudenmaier, *op. cit.* (2), 95–103; Vincenti, *op. cit.* (17), 3–6.

52. 'Watt's discovery [his "steam engine"] was no accident. Nor was it an accident that the discovery had an overwhelming application in the field of production.' Birdsall and Cipolla, *op. cit.* (49), 158.

53. Cf. my piece, 'A Cambridge View of Modernity', *op. cit.* (45).

54. A. Koestler, *The Act of Creation* (London, 1964); *History, op. cit.* (4), 1–2.

55. Gellner. *op. cit.* (3), 16–17, 21.

56. *Ibid.*, 20. 'Mankind has passed through three fundamental ecological stages: hunting/gathering, agriculture, and industry.' *Ibid.*, 275.

57. M.P. Gilmore, *The World of Humanism 1453–1517* (New York, 1952), being part of the Harper series edited by W.L. Langer.

58. Cf. the title of S. Giedion, *The Eternal Present: The Beginnings of Art* (Oxford, 1962).

59. J.H. Plumb, *The Horizon Book of the Renaissance* (New York, 1961), 92–3; K. Clark, *Civilisation* (London, 1969), 98–9. In Clark's book the painter of Ideal Town/City has been identified as Piero della Francesco, though the precise human agency is irrelevant here. Note too that the editors of *The Penguin Book of the Renaissance* (Harmondsworth, 1991), based on the 1961 volume, have promoted the display of an empty Ideal Town/City onto the volume's front cover. 'The Renaissance without People. Discuss.'

60. E.g. Clark, *op. cit.* (59), 114–15, where reproductions of two paintings by Giorgione have been displayed. 'With Giorgione's picnic the balance and enjoyment of our human faculties seems to achieve perfection' (p. 114).

61. This theme has been set out more fully in M. Fores and R. Porter, 'Modernity and Apocalypse Then', typescript, Wellcome Institute for the History of Medicine.

62. S. Shapin, *The Scientific Revolution* (Chicago, 1996), 13.

63. Gellner, *op. cit.* (3), 64, 65, 66 in ch. 2, 'Community to Society'. Note the use of a metaphor implying falsely how all this cultural newness was taking place on a different planet from that of earlier cultural experience.

64. *Ibid.*, ch. 2. The last two sections of this chapter are 'The Terminus' (pp. 62–7) and 'The Overall Plot' (pp. 67–9).

65. *Ibid.*, 68. Compare the first sentences of ch. 5, 'Codification' at p. 113: 'The Enlightenment codified the vision of the world as a unified, self-sufficient Nature, governed by orderly laws, and including man within itself as a part of Nature and nothing else. It is the codification, in outline, of the world we now inhabit.'

66. Cf. the title of Gellner, *op. cit.* (3), ch. 4, 'The Tension'.

67. *Ibid.*, 92, near the start of ch. 4.

68. *Ibid.*, 128, 127, from ch. 5, 'Codification'. Note the dramatist's use again of an apparent assumption that two separate planets have been in the frame.

69. Whitehead claimed that the 'modern' secular belief in 'natural' order can only have derived from the belief in 'the rationality of God' of medieval Europe, a belief in the agent of making so much order in the first place. *Op. cit.* (44), ch. 1.

70. Cf. the discussion of *Plough*'s chs 4, 'The Tension' and 5, 'Codification'.

71. See also the argument of 'Introduction: Mystery about a Base', in Fores, 'Hamlet', *op. cit.* (9), 161–6.

72. M. Fores, 'Newton on a Horse', *History of Science*, 1985, 23: 351–78.

73. Birdsall and Cipolla, *op. cit.* (49), 36, 158.

74. R.S. Westfall, *Never at Rest* (Cambridge, 1980), 58.

75. Cf. Fores, 'Hamlet', *op. cit.* (9), 165, on how 'much of the latter-day, generalized treatment of the "relationship between science and technology", two ill-defined entities, has only related to events in a meta-world separate from the single world of post-Newtonian conception.'

76. Fores, 'A Cambridge View', *op. cit.* (45), 227–9, has had him more at home in the tea-room of the University Library in Cambridge, England, than in deepest 'industry'.

77. M. Fores, 'Francis Bacon and the Myth of Industrial Science', *History of Technology* 1982, 7: 57–75, esp. pp. 71–3. See also some other examples of the effect in Fores, 'Hamlet', *op. cit.* (9), 168–82.
78. Cf. note 1 above.
79. Bacon, *op. cit.* (1), Bk 1, Aphorism 70.
80. *Ibid.*, Bk 1, Aphorisms 46, 84.
81. *Ibid.*, Bk 2, Aphorism 41.

John Farey Jr (1791–1851): Engineer and Polymath

A. P. WOOLRICH

In the Royal Scottish Museum, Edinburgh, is a boxed set of engineer's drawing instruments, presented to John Farey Jr in 1844 by Robert Stephenson. Accompanying them is a manuscript biographical note of Farey by his only child, Edward (1825–1913), which reads:

> John Farey was born in Lambeth 20 March 1791 and was brought up at Woburn the native place of his father who was agent to Francis, Duke of Bedford. He was engaged at the early age of 14 years and for some years after drawing for Dr Rees' Encyclopaedia, Tilloch's Mag, Gregory's Mechanics and Mechanical Dictionary, Edinburgh Encyclopaedia, Pantologia, Smeaton's Reports and other Scientific publications. He also wrote many articles for these works.
>
> In 1806 when aged 15 years of age he was elected a member of the Society of Arts and became acquainted with many scientific men amongst whom Mr Telford who presented him with Smeaton's original papers & drawings — Also Sir Joseph Banks, Mr Bramah, Brunell. In 1807 he invented his vanishing point instrument, published in the Society (of Arts etc) transactions., Vol 32, page 71 and for which he received their silver medal in 1814. In 1813 he received Gold Medal from Society of Arts for an instrument for drawing ellipses.
>
> In 1813 he made an instrument for drilling barrels of musical boxes and chimes and made a musical box, believed to be the first made in England and a chime clock, he also set out the patterns for several drawing instruments.
>
> In 1819 he went to St Petersburg and other parts of Russia and while there he found an indicator, he experimented with it to discover its use and it is believed he set McNaught to make them on his return.
>
> In 1821 John Farey commenced writing his work on the Steam Engine and determined to give up his profession of Civil Engineer in favour of his brother Joseph, having made arrangements to enter a firm of lace manufacturers for which purpose he resided in Devonshire until 1823 (Decr), when he gave it up.
>
> In 1825 he went to reside in Leeds to manage the mills of Messrs Marshalls, but returned to London in 1826 and continued the

> profession of engineer as a consequence of his brother Joseph becoming paralytic. He continued busily engaged in most of the new discoveries and scientific subjects until his death 15 [*sic*] July 1851.
>
> He was gifted with universal talents and all subjects seemed to be familiar to him, were they Engineering, Chemistry, Astronomy, Mathematics, Geography, Geology, or Theology, on which subject he at one time wrote considerably. In the early part of his life he was proficient in music, singing, painting and engraving but more important subjects prevented his continuing them.

The engraved plaque on the drawing instrument box simply records 'To John Farey, Esq, from Robert Stephenson'.

Edward's biographical note of his father is a very odd document, considering what has been left out. The narrative up to 1826 is unexceptional, notwithstanding some inaccuracies of fact, but much detail of Farey's working life is ignored, and there is no mention of his brothers (other than Joseph) and sisters. There is nothing about the death of John's first wife (Edward's mother) in 1825, and his remarriage in 1827. After 1826 the notes are skimpy. They ignore his material contribution to many parliamentary investigations such as the one on patent law of 1829, and there is no mention of his widespread involvement with the patents for numerous important inventions of the time, especially those involving textile manufacture. Also ignored is his work as a consulting engineer to men like John Heathcoat, the inventor of lace-making machinery, and Cook and Wheatstone, the inventors of the electric telegraph. His links with France and the United States do not figure, and above all the dreadful fire at his house in December 1844 in which four people were killed is not mentioned. All this is described as 'continuing busily engaged ...'! It is strange how Edward simply ignores it, but it is roughly the same emphasis as the *Dictionary of National Biography* gave, which in turn follows the memoir published by the Institution of Civil Engineers in 1852.

John Farey Jr was the eldest son of John Farey Sr, the land surveyor and geologist, and Sophia Hubert, and was born in Lambeth on 20 March 1791. A year later the family moved to Pottsgrove, near Woburn, Bedfordshire, following the appointment of John Sr as land agent to Francis, 5th Duke of Bedford. Full details of the Farey family will be found in the Appendix to this paper.

Nothing is known of the early education of John Jr, although it is clear that he received much informal education from his father, who took him, as a very young child, to attend lectures given by itinerant teachers, perhaps at Leighton Buzzard.[1]

Nothing has so far been discovered to show where John Jr was educated after the family moved back to London in 1802, but it is clear he must have picked up many scientific and technical insights from his father's circle of friends and clients.[2]

James Harrison suggests that John Farey Jr may have had training in patent work from William Nicholson (1753–1815), the technical writer and patent specification draughtsman, who also ran a school at his home in Soho Square, London. Certainly Farey and Nicholson did have close links, since Farey drew many of the plates illustrating Nicholson's *British Cyclopaedia* published in 1809, and in 1811 they acted jointly in producing the specification of Joseph Chessborough Dyer's wool-card making machine (no. 3498). In addition there is evidence that 'John Farey' had a hand in a patent trial which took place in 1802, but this was probably John Sr since it is not likely that the 11-year-old John Jr could have become involved in a court case in his own right.[3]

The formal education of John Jr seems to have ceased by 1804.[4] In that year he began the systematic study of industrial machinery in the London area. He spent his entire time for the next two years on the project, which involved visiting mills and machine manufactories, taking detailed sketches. Two of his notebooks survive with notes of steam engines, textile machines, lathes and a number of sketches of named mills showing how the machinery was positioned (Figures 1, 2 and 3).[5] These are not fully worked up finished drawings, however, and none such has been discovered from this period of his life.

The young Farey developed a precocious talent for mechanical draughtsmanship, as did his brothers and sisters. Later all the family assisted in the business. Farey illustrated and contributed to periodicals and technical books and also multi-volume encyclopaedias. He is also known to have written a number of technical books pseudonymously. Much of his writing for encyclopaedias involved recycling information from a multitude of sources, and the question arises as to whether this might be regarded as plagiarism. It is very hard to answer this since copyright law then was different from that of today, and modern views of plagiarism do not necessarily apply. Also, there is a difference between literary writing and technical journalism, and the conventions for referencing then were quite different. It was common for a writer simply to cite the name of the author, and perhaps volume number and page, leaving it to the knowledge of the reader to fill in the details of title and place and date of publication. This can give the impression today that the writer was not over-scrupulous in citing his sources.

Farey designed various instruments to assist in the production of his drawings, which he sold to fellow artists.[6]

As well as acting for publishers the family undertook much technical work for clients, especially in regard to patents and inventions. Henry Farey, a younger brother of John Jr, kept a diary, for which the volumes covering the years 1816, 1817, 1819 and 1820 survive, and these give a marvellous insight into the work the family did, and a unique portrait of the world of the Fareys and their friends. Henry, who was aged 15 at the beginning of 1816, ran errands and copied drawings, but as he got older he undertook more work of his own, and the diary for 1820 ends with him working in Lancashire cotton-mills, drawing spinning and weaving machinery. All the diaries are filled with minute sketches and scenes of the Farey household.[7]

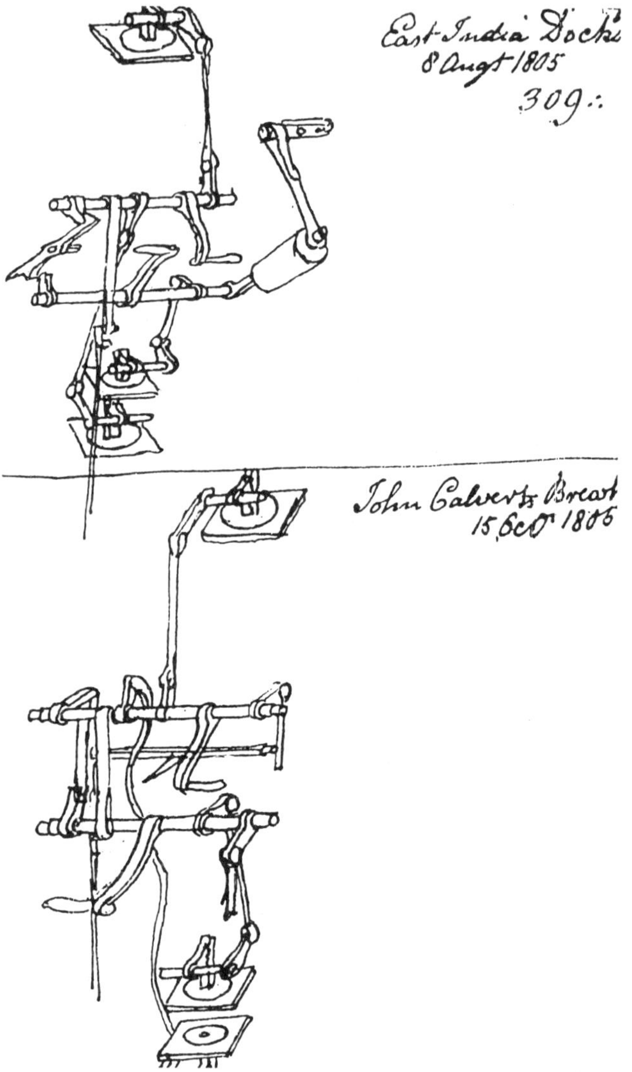

Figure 1 Perspective sketches of steam engine valve-gears. By courtesy of the late Arthur and Edith Farey.

A detailed account of the technical work the family did is beyond the scope of this paper, but nevertheless it must be touched upon briefly. In 1813 John Farey (probably John Sr) was involved with Aaron Manby and Samuel Clegg in the formation of the Gas, Light and Coke Company, to provide coal gas illumination for the metropolis.[8] In 1814 John Farey Jr was approached by James Watt about the specification of a sculpture-copying machine which Watt was developing. This link ended only with Watt's death in 1819, for he got into the habit of waiting on the Fareys to

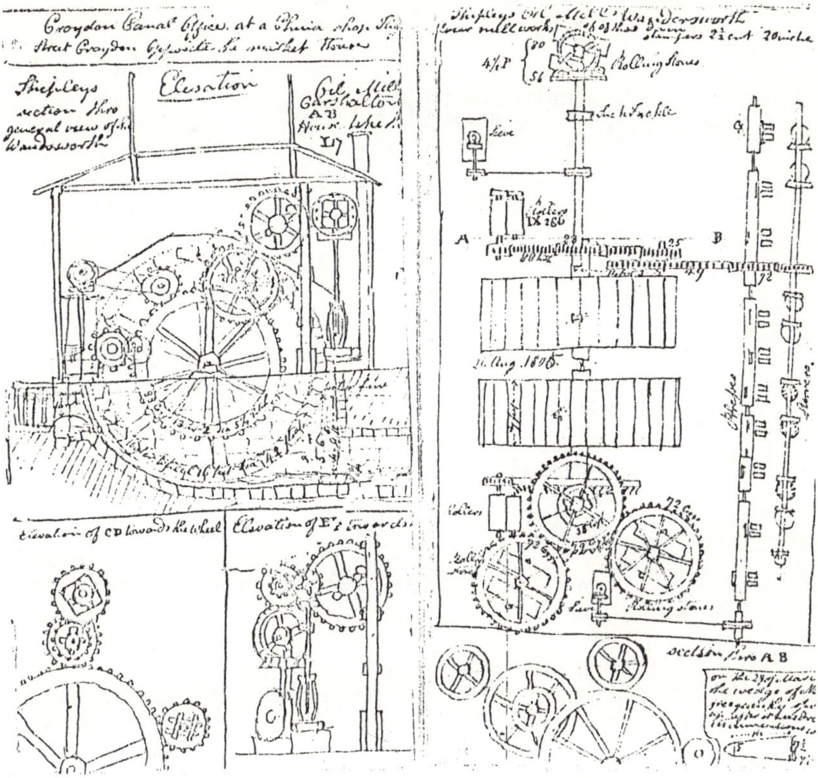

Figure 2 Sketches of layout of Shipley's oil mill, Wandsworth. By courtesy of the late Arthur and Edith Farey.

gossip about the latest technical developments and new patents whenever he went to London.[9]

One of the more important clients of John Farey Jr was John Heathcoat (1783–1861), the inventor of lace-making machinery, and over the years the family did a great deal of business with him, including building models used in patent litigation.

The family also had very close links with France and French engineers. In September 1814, John Farey Jr visited Paris and sketched the waterworks steam-pumps at Chaillot, built by Pèrier Frères. In 1816 Pierre François Montgolfier, son of Joseph Michel Montgolfier (1740–1810), who invented the hot-air balloon and improved the hydraulic ram, stayed with the Farey family while John and Henry built the prototype and arranged for the British patent of Montgolfier's improved hydraulic ram (no. 3992) and his expansion water lifter (no. 3995). John Farey Jr was also active in France for British engineers who wished to take out patents there, such as Heathcoat and the sons of William Eaton, the inventor of the mule jenny (no. 4272).[10]

The family also had close links with Philip-Joachim Gengembre

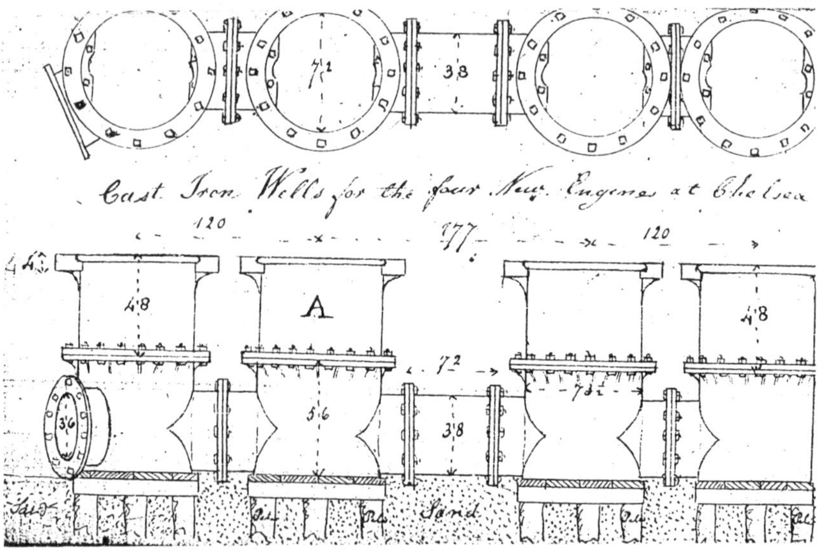

Figure 3 Cast iron wells at Chelsea waterworks. By courtesy of the late Arthur and Edith Farey.

(*c.* 1763–1838). Gengembre was an important engineer, regarded as the father of the French steam navy. He was originally trained as a chemist under Berthollet and was involved in the discovery of phosphine. He went to America during the French Revolution, and on his return to France secured a position at the Paris Mint. On the return of the Bourbons he lost his post, and then travelled to England where he visited the Farey family, in 1818. He was accompanied by his son, Charles Antoine Colomb (*c.* 1790–1863), an architect, and the young man's interest in the Farey family quickly became something more than professional, for on 12 April 1818 he was married to Marianne Farey at St Pancras Church, London.[11]

John Jr visited France several times on Heathcoat's behalf. In 1819 he spent two months in Russia. The reason for his visit is not clear, but he later stated that he saw the St Petersburg Mint, supplied with machinery by Boulton & Watt but erected by Scottish engineers (probably supplied by the Carron Company).[12] During this trip he saw an ironworks, possibly that erected *c.* 1770 at Kolpena on the Neva by the Carron Company. He also inspected the Kronstadt dockyards, where he saw the atmospheric steam engine designed by Smeaton in 1774. He discovered details of the steam engine indicator of James Watt, used by the firm's erectors to tune their engines, and on his return to Britain was responsible for its introduction generally.[13] When he returned he landed at Leith, a short distance from Carron, where he visited the works, presumably to report on his observations. He remained in Scotland for a month before returning to London.[14]

Giving evidence to a parliamentary committee investigating the provision of small arms for the army, which sat in 1854, Richard Prosser said that:

> about 1822 Mr Fairey [*sic*], engineer of London, went to Russia and at Toola [*sic*] he saw his machinery [gun-making machinery of John Jones] in operation. It was some time about 1817 that it went out and he saw there 12 Russian soldiers come with muskets made by this machinery.[15]

Whether Farey saw this in 1819, or made a subsequent visit to Russia, is not known. The latter is more probable, since Tula is more than 500 miles from St Petersburg, and the four weeks Henry's diary records him as being away from Britain does not seem to allow enough time to get in all he is known to have done, and add in a more than 1000 mile round trip to see Tula.

John Farey Jr had the reputation of being a meticulous researcher for precedent in his patent work, but must have experienced difficulties both from the problem of obtaining information and from the growth of the number of patents with which he had to be familiar.

The British patent system, which began in 1617, had degenerated into a cumbersome, time-wasting and expensive bureaucratic machine, largely incapable of serving the needs of the inventor. Each patent adviser needed to compile elaborate collections of information himself, since the Patent Office library did not exist before 1856, while original patent documents were split among three possible repositories, the Enrolment Office of Chancery, the Petty Bag Office and the Rolls Chapel Office, at the option of the inventor, and public indexes of the documents were not available. New patents were described in the technical and scientific press, but not systematically, and a man like John Farey, able to offer good and accurate advice to inventors, and also to capitalists who wished to exploit inventions, was assured of constant work, and was able to charge high fees.

Patent law and its administration at this time were a minefield for the unwary, and there was much agitation for reform – which did not in fact take place to any degree until after Farey's death in 1851. The major area of difficulty was the cumbersome legal manoeuvring the patentee had to go through to get his invention patented and the many fees he had to pay along the way, as well as the fees he had to pay his legal and technical advisers for getting the specification of the patent drawn up and the drawings prepared in the first place. In 1829 an investigation was carried out by a Parliamentary Select Committee before which Farey gave a great deal of written and verbal evidence. He showed how a patent was obtained, gave detail of the abuses an inventor was likely to experience and set out his views on how matters might be improved. He summed up his opinions in a letter to Macvey Napier, Editor of the *Edinburgh Review*, written on 28 April 1830:

> The present system is admirably calculated to increase the number of patentees for absurd inventions who pay their fees whilst they are under the delusions of perpetual motions, gaining powers, raising water without cost and other mysteries, which vanish when exposed to the light after the patent has been granted. A real hotbed for

> distempered imaginations. The fees of Office are thus augmented, but the continual failure of patent rights when brought before the courts of law and the ruin of inventors deters all our best and soundest artists from expending their money on the faith of protection by law, and such men would probably do nearly as much as they do now if the system of promising protection (or rather taking money for such promise which is scarcely ever fulfilled) were entirely abolished.[16]

According to the statistics of patents (compiled by Benett Woodcroft when the old law was changed in 1852), the annual grant of English patents rose from 95 in 1805, when Farey began, to 455 in 1851, the year of his death.[17] In practical terms this meant that Farey, acting as an adviser, needed to know about 2500 patents already granted when he commenced practice, a number which had climbed to a total of around 14,000 when his career ceased. To these must be added figures for Scottish and Irish patents as well as those for foreign countries. It was forbidden for an inventor to take notes from the original patent rolls, and if he wanted to know in detail what his predecessors had done he had to pay high fees to the Crown legal officers for the documents to be copied officially. The patents were handwritten in law-script onto parchment skins which were sewn end to end before being rolled up for storage; it was thus a very cumbersome business to obtain copies with any efficiency. Farey must have set out with the intention of accumulating technical information as a service to his patent clients, and found he could make use of it over and over again in his writing and drawing activities.

Farey long continued writing articles for Rees's *Cyclopaedia*, and on 26 October 1816 was published the half-volume containing his article on the steam engine.[18] It was well received by his fellow engineers, who urged him to produce a full-length work on the subject.[19] He had, in fact, been accumulating information for many years for a far larger project, a series of books about technological methods, as he described in a letter to Thomas Coates, of the Society for the Diffusion of Useful Knowledge, written in May 1830:

> More than twenty years ago [i.e. before 1810] I formed a plan of an extensive series of publications, on the subject of my professional pursuits, and have never omitted any opportunity of collecting materials for its execution, when I have more leisure than professional business afterwards allowed me. I have composed many detached articles for it. Fragments have appeared in Rees Cyclopaedia, The Edinburgh [Encyclopaedia] and many other miscellaneous publications.[20]

Farey commenced writing the text of *The Treatise on the Steam Engine* in 1820, and in the same year the bulk of the 60-odd drawings was produced. In 1821 he gave up his interest in the family business in favour of his brother Joseph, and the concern was run from an office at 44 Lincoln's Inn Fields, London. John moved to Devon, where he became involved in

Heathcoat's lace manufactory at Tiverton.[21] When he moved from his father's house in Howland Street, John dropped 'Jr' from his name when writing letters and articles.[22]

On 1 February 1822, John Farey was married, at St Pancras Church, London, to Mary Horniman Thompson Taylor. There are a few entries to the Misses Taylor in Henry's diaries, but nothing more has so far been found about Mary's parentage.[23]

Farey's activities in Devon are obscure, but it is plain that he assisted in developing Heathcoat's improved lace-making machine, patented in 1824 (no. 4918). A curious feature of Farey's involvement with Heathcoat is that it is ignored in Felkin's *History of Machine-Wrought Lace* (1867). Felkin was also an employee of Heathcoat at the time Farey was working for him, and it is quite impossible to believe that Felkin was ignorant of the facts of Farey's role in the development of the inventions. The only place where Farey is mentioned in Felkin's book relates to his giving evidence at the 1816 patent trial. What Felkin's motives were in distorting history in this way can now only be guessed at. It is possible that personal animosity may have played a part, but it is more likely to have been an example of a common failing of Victorian hero-worshipping writers, a wish to believe that the men they wrote about did it all single-handed, and that no other persons had any part in their success.

Farey and his wife returned to London in December 1823, and he restarted his practice as a consulting engineer. In the spring of 1824 Mary Farey became pregnant and in the summer John was consulted by Marshalls, the Leeds flax spinners, about improvement to their mill. This involved designing, building and installing the new plant and so entailed moving the family to Yorkshire. Nothing has so far been found about the detail of Farey's work there but it must have involved the kind of relationship he had with Heathcoat and which he had in later years with other capitalists.[24]

On 15 February 1825 Mary Farey gave birth in London to a son, Edward (1825–1913),[25] but she died eight days later.[26] Farey appears to have returned to his duties in Leeds, but in the summer of that year his brother Joseph was taken ill and became paralysed.[27] John Farey Sr took over the business for the rest of the year, but he too died on 11 January 1826, and was buried at St James, Piccadilly, London. There was also the accidental death on 2 May 1826 of his cousin James Cundy, the sculptor, who was killed by a furiously driven butcher's cart in Piccadilly.[28] This run of events was probably the reason why Farey gave up his work with Marshalls and returned to London, for not only did he have to run the business but also he would have become responsible for his mother and unmarried sisters in addition to his infant son, who was sickly.

By October 1826, the text of the first volume of *The Treatise* was almost ready for publication and in January 1827 Longmans, the publisher, wrote to Farey the first of a series of letters pressing him to get on with finishing the second volume. Only the Longmans part of the correspondence survives, but from this it is plain that Farey was a very tardy author, though in fairness it must be said that he had been through a very trying time, and when Longmans began writing he was in the middle of preparations for his second marriage.

His new wife was Elizabeth Pugsley (1791–1850), the daughter of Robert Pugsley of Ilfracombe in Devon. How they came to meet is not known. Perhaps it was while he was working for Heathcoat at Tiverton, for her family were living there at the same time. Alternatively it might have been through her brother-in-law, Edward Holroyd, in 1826 a young lawyer who in 1830 published a book on patent law and who may have had professional dealings with Farey. Holroyd's brother was brother-in-law to William Morgan, the marine engineer, and inventor of the feathering paddle wheel, who may also have had dealings with Farey, so perhaps the link lay there.[29] The marriage took place on 24 February 1827 at St Pancras Church, London.

Details of Farey's professional life for the next two years are scrappy, but Longmans kept pressing unsuccessfully for him to complete the second volume, and in the summer of 1829 took legal proceedings against him. Although the project was not entirely abandoned, the second volume, as originally conceived, was never written; instead a volume was produced about high-pressure steam and the Woolf pumping engine, but it was not published in Farey's lifetime.

There had been extenuating circumstances for this further delay, for in April 1829 Farey had been giving evidence to the Parliamentary Committee on the law relating to patents for inventions, and must have spent a considerable amount of time assembling his material. In July 1829 Joseph Farey died at his brother-in-law's house in Paris, from blood poisoning caused by a cut behind the ear from a barber's infected razor.[30]

Then Farey himself then fell ill for about nine months from the middle of 1829, and early in 1830 his mother died; he was himself ill again in July of that year. Early in 1830 Farey had been approached by the Society for the Diffusion of Useful Knowledge to write something for the society about the steam engine, and in his reply, of May 1830, Farey commented about his problems with *The Treatise*:

> My volume on the Steam Engine is the only part of my plan which I have so far made complete as to desire to publish it in my own name, but I find the labour of rendering a work so exact as to instruct professors, and to be a safe guide for students towards real business, is too great to be continued, at least on that scale and the publishers find the sale too limited. The wreck of my project will furnish some useful articles ...[31]

In June and July 1830 Farey submitted affidavits for the plaintiffs in the famous patent trial *Cochrane and Galloway* v. *Braithwaite and Ericsson*. This case concerned steam boiler designs. There are signs that something was wrong with Farey's judgement, for the second affidavit, heard on 31 July 1830, in some measure contradicts opinions expressed in the first, heard on

30 June and 6 July 1830. The editor of *Mechanics Magazine* commended Farey for altering his opinion, but the question arises as to why he gave a different one to begin with. Perhaps the answer lies in the state of his health.[32] The nature of Farey's illness is unknown, but it appears to have induced forgetfulness; he mentioned this in another letter to Coates, written on 23 July 1831:

> I am obliged for the copy of your letter you sent me on 19 July 1830: on reading it I recollect having had such a letter read to me at a time when I was too ill to sit up. I was obliged to leave town shortly afterwards for the recovery of my health, and at return it had totally escaped my memory that I had ever received such a letter.[33]

Sometime between January and June 1833, Farey moved to a rented house and office at 67 Guilford Street, Russell Square, London, a large property containing sixteen rooms on five storeys, with outbuildings and stables to the rear.[34] The Farey household comprised John and his wife and son, and a small number of servants, as well as an apprentice, or clerk, and a coachman who lived nearby. Farey seems to have organized his office as a family affair, just as it had been in his father's time at Howland Street. He was assisted by his wife, who must have been an exceptional lady, described as

> a lady of rare attainment, who to great amiability of disposition, and a thorough knowledge of the duties of her sex, joined by almost masculine scientific attainments which she employed as modestly as profitably in aiding the labours of her husband.[35]

Farey's neighbours were professional men like himself, mainly lawyers, doctors and artists. Many had connections with the nearby Foundling Hospital. Also at 91 Guildford Street, from 1830 to 1834, lived his brother-in-law Edward Holroyd and his family. By this time Edward had been appointed a Commissioner for Bankrupts and seems to have given up his connection with patents advocacy, though in 1830 he had published a book on the law of patents.[36]

Joseph and John were elected Members of the Institution of Civil Engineers in 1822 and 1826 respectively, and in 1834 John was elected to the council, where he urged reforms of the constitution and rules, eventually carried out in 1837.[37] Among the more important reforms were those relating to education and training of engineers. Farey was very much in favour of the traditional British method of apprenticeship or pupillage, as opposed to the Continental method of academic training in special schools, such as the École des Mines. Farey must have been one of the few practising engineers of the time who had close familiarity with European custom, and the later progress of British technical education might well have followed a very different route had he been in favour of the Continental system and been able to persuade the Institution to support it. When later technical institutions came to be formed they modelled themselves closely on the existing structure of the Institution of Civil Engineers.

Farey's professional work continued much as before, but he seems to have dropped much of his writing and drawing activities for publishers in favour of engineering consultancy. He was a witness to various government inquiries including the Parliamentary Committee on Steam Carriages of 1832 and the Lords Committee on Patent Law Reform of 1835. He also appeared at inquests such as the one held to investigate the *Victoria* steamboat boiler explosion of July 1838.[38] He reported in 1832 (with his friend Joshua Field) on the relative merits of round and square fireboxes in locomotive boilers for the directors of the Liverpool and Manchester Railway.[39] Farey's patent work continued. He was involved with the Erard family in their piano patents[40] and with Robert Stephenson's long-boilered locomotive,[41] and he acted for several of the inventors who introduced steam road carriages, such as Walter Hancock, Sir Goldsworthy Gurney and the Heaton Brothers of Birmingham. His main activity seemed to have been in textile machinery, and he was concerned in drawing up the longest patent specification drawn up under the old law. This was for William Croft's 'Improvements in certain machinery for making figured or ornamented bobbinet ...' (no. 6854), which was enrolled on 26 December 1835. William Croft, of New Radford, Nottingham, was employed by Fishers, the lace manufacturer. He made a series of improvements during the 1830s, for which Farey drew up the specifications. This involved Farey working at Nottingham, presumably in the same way he had been employed by Heathcoat at Tiverton and by Marshall at Leeds a decade earlier. The monster patent incorporated nine distinct inventions. The specification used 65 skins of parchment and ran to 1168 handwritten folios. There were also 49 sheets of complex drawings. Fishers paid £117 in stamp duty alone, and the firm's total costs were said to have amounted to between £4000 and £5000, of which a large proportion must have been Farey's fee.[42] The length and minuteness of Farey's technical description in the specification was censured by the editor of *Mechanics Magazine*, who concluded:

> It is, indeed, not wonderful that the patent law, originated without principle, and proceeding only by blunders, should involve even well-meaning men in absurdities, which bring on them the suspicion of caring most for their professional gains .... It is not for us to allot between Mr F. and the law their respective shares in the monstrosity of this notable specification. ... The mischiefs consequent on the present law might certainly in this case have been greatly diminished had Mr F. been less attentive to the amount to which he was swelling his client's costs.[43]

The main complaint of *Mechanics Magazine* was that the specification comprised nine inventions, and since the rules prevented rival patentees from taking notes from the original patent, they were faced with paying around £40 for the entire specification to be officially copied and at least £100 more for the drawings. Since only the complete document was permitted to be copied, a man who was concerned with only one section was forced to waste his money on the sections which did not concern him, and so the patent was an oppressive financial burden for the small improver.

This was not the first time that Farey had been criticized by the editor of *Mechanics Magazine*, for the journal was somewhat rude about his evidence to the 1835 Lords Committee on Patent Law Reform. Farey was described as a 'consulting engineer of considerable standing who went prepared to occupy the committee for a couple of days with his crotchets and speculations, but who was civilly dismissed after a few questions asked and answered and who, it must be confessed, was likely to have been more verbose than either pertinent or edifying'. The committee met in secret and the evidence was never published, so it is hard to see the basis for these comments.[44] The editor of the *Mechanics Magazine* was Joseph Clinton Robertson, also a patent agent, so perhaps his remarks were fuelled by professional jealousy.

One of Farey's bills was disputed by Fishers, who claimed that he charged them for correcting his own errors in the specifications, and this got as far as Farey suing Fishers.[45]

Farey also drew up the specification for Cooke and Wheatstone's electric telegraph of 1837 (no. 7390). Cooke described Farey as 'the best consulting engineer in England',[46] but a modern commentator has criticized the patent for being badly drawn up, and says 'it is as if it was written by the patent agent with the two patentees leaning one over either shoulder and dictating simultaneously into either ear. Perhaps it was. At least it was drawn up in a great hurry.'[47]

Farey sometimes travelled to France to visit his relatives, and in 1842 a daguerreotype (Figure 4) was taken of him by his nephew Henri Gengembre at the family home at Indret.[48]

In the spring of 1842 Farey took part in two discussions at the Institution of Civil Engineers about the steam engine indicator. In the second of these, held on 19 April 1842, he spoke about indicator design and displayed a series of indicator cards from a range of engines. He was often called in to make tests of engines with a view to discovering their efficiency.[49]

In August 1842 Farey wrote with his own hand his will, in which he left his widow the household effects and his son Edward his technical library and workshop equipment used in his professional work.[50] Farey kept his papers in iron fireproof chests and was fully conscious of the value of his collection, since he willed that it was not to be dispersed but kept together by his son for his own use as a consulting engineer. Farey made elaborate plans for catalogues to be prepared of both bequests in case of dispute. Copies of them were to be kept by both beneficiaries, and also by Farey's executor, Joshua Field, of Maudslay, Son & Field.

During the first half of 1844 Farey was in Ireland and spent some time examining the Dalkey and Kingstown atmospheric railway; he spoke about it in May of that year in a discussion following a paper given to the

Figure 4 John Farey at age 52. Daguerreotype made in France, 1842, by a member of the Gengembre family. By courtesy of the late Arthur and Edith Farey.

Institution of Civil Engineers.[51] In June he displayed his elliptograph to a conversazione held at the Institution, and on 16 December was a witness at the important patent trial *Lowe* v. *Penn*. This concerned the invention of the screw propeller.[52] Also in December of that year he was a witness to the Parliamentary Committee investigating the state of large towns and populous districts, where he presented evidence about the efficiency of various steam pumps used for supplying water.[53]

In the early hours of Saturday, 21 December 1844, a disastrous fire occurred at Farey's residence. The family with their servants were able to escape over the roof to the next-door property, though Farey's 15-year-old clerk, William Robinson, was burned to death, together with two policemen named Berkmeyer and Wright who had entered the basement to restrain a neighbouring coachman who had entered to rescue the Fareys' dog, and who also died. The smoke was so dense and the noise of the fire so great that the cries for help made by the family were unheeded by the crowds below, as Farey described to the inquest:

> If such an awful scene could be called ridiculous it was to see us all on the housetop, looking down into the street, with multitudes below, and shouting as loud as we could, and striving by every means in our power to attract notice, and not a person seeing us or attempting to assist us. We got ourselves out into the next house window. We could get no assistance, but were on the housetop for a quarter of an hour or more naked. I broke the nails from my fingers in endeavouring to tear the slates from the roof to throw down to the police in order to attract their attention, and knocked and thumped at the skylight of the next house to wake the inhabitants, till the skin was off my hands and elbows, but no one came or seemed to notice us for at least a quarter of an hour.[54]

They were seen eventually, and the police broke into the next-door house and went to the roof to rescue them.

The fire was centred at first on the first floor, under Farey's study, but quickly spread to the floors above. The house still stands, and the brickwork facing Guilford Street still shows signs of blackening by smoke (November 1990). Colonnade Mews to the rear is little altered, and it is not hard to image the neighbourhood as it was in Farey's time.

The collapse of the upper floors fractured cast iron beams let into the second floor which supported the weight of the fireproof chests, and this mass of burning debris descended into the basement, causing three of the deaths; the collapse seems to have had the effect of snuffing out many of the flames. The fire was caused by the ignition of soot in a blocked flue pipe. Farey's house had been altered by a previous tenant, John Sims, FRS (1749–1831), with the result that the *trompe-l'œil* pillars in the drawing-room were incapable of taking any weight.

The Fareys and their servants were sheltered in the house of the Martineau family, who lived opposite at no. 55. They were related to the partner of Philip Taylor of the engineering firm of Taylor & Martineau, and like him were Unitarians. Within a few days the Fareys moved to a property at Regent Square, a few streets away, where John set up his office for a few months.

In the weeks after the fire, efforts were made to salvage what had survived, and from the newspaper reports it is plain that a considerable proportion of Farey's collection was saved, little harmed. *The Times* reported:

> In the process they found Mr Farey's two iron fireproof safes, which are of immense weight and which indeed materially contributed to the falling of the floors and the consequent loss of life; to get them out required the united exertions of twenty men. The contents which consisted of valuable plates and other property were found uninjured. An immense quantity of books, models etc., were also got out which are stated to be of extreme value, and were conveyed to the new residence of Mr Farey in Regent Square. Many of these productions have an important bearing on science and could not have been replaced if lost; some of them were the result of forty years research, experiment and labour.[55]

The *Observer* report, although garbled, contains more clues about what was salvaged:

> In turning over the immense pile of burnt rafters etc., a number of valuable models, connected with machinery have been recovered, besides a quantity of scientific works, which, though much damaged are not [so] extensively injured as to be rendered valueless to professors of Science. Amongst these are nearly 300 volumes of Knox's celebrated works upon Smeaton, and perhaps not the least prized relic that has been saved is a sliding side rule for measuring cylindrical steam boilers etc., presented by the renowned Mr Watt, of Bolton, Lancashire, many years since to Mr Farey sen. [i.e. John Farey Jr]. Two ponderous iron safes filled with works of art have been removed from the basement of the ruins.[56]

'Knox' is perhaps a misreading of a shorthand note about 'Banks' – Sir Joseph Banks, from whom Farey obtained the Smeaton MSS at the beginning of his career. The 'sliding side rule' presented to Farey by 'Mr Watt, of Bolton' (*sic*) was a Soho-pattern slide-rule which had been developed by Watt and his assistant Southern, and which was used by the firm's staff when making calculations for engine designs. A Soho-pattern slide-rule, perhaps this one, is in the possession of the Farey family in the United States.

Farey's property was insured by the Sun Fire Insurance Company for the sum of £1500, but the newspapers commented that this was unlikely to cover his costs. Nothing more has been discovered about the extent of the damage, but Farey was quickly able to continue his professional work, which would not have been possible if all his books and records had perished in the blaze. Farey moved to 35 Guilford Street by mid-1845 and remained there for about a year while no. 67 was rebuilt.

In February 1847 he was proposed for Fellowship of the Royal Society, but was blackballed.[57] The preprinted Certificate of a Candidate for Election is completed in Farey's handwriting except for the signatures of the sponsors. He described himself as the author of 'several detached articles and numerous drawings in Rees' *Cyclopaedia*, Brewster's *Encyclopaedia*, and Napier's *Supplement to the Encyclopaedia Britannica*, the drawings for the publication of Smeaton's *Reports*, (1806–1820), *A Treatise on the Steam Engine*, 1827' and the inventor or improver of '[t]he Elliptograph for describing ellipses and dividing their circumference, for perspective drawings and engraving'. He was accomplished 'for his acquaintance with the science of Mechanics in all its various branches [and as a] Consulting engineer, in matters relating to the applications of science to practical purposes'. Taking Farey's modest description of himself at face value tells nothing at all about the work he really did. Surely he ought to have made much more of the projects he was involved with as a consulting engineer. Was the list of his writings and only one of his inventions some 25–30 years previously of any relevance when this certificate was written? The list of his sponsors is very impressive on paper, but there is no way of knowing whether these names reflect fairly the social and political balance of the Royal Society as a whole.[58] The majority were elected to the Royal Society in the 1820s and early 1830s, and some had effectively retired from practical matters. Were Farey's sponsors from a wing which was then out of step with the rest of the membership? The fact that Farey was blackballed makes it appear that they were.

In 1847, 1848 and 1850 he took part in discussions following lectures to the Institution of Civil Engineers on fireproofing buildings and also on locks and safe construction. In none did he make any reference to his own experiences, apart from the comment that the growth of fire insurance tended to make people too careless about the danger of fire.[59]

His consultancy work continued. In 1848 he was engaged by an American, Isaac Babbitt, to carry out tests, with some other engineers, on Babbitt's bearing-lining metal, and their findings were published by Babbitt in that year.[60]

In 1849 Farey conducted trials on the Bischopp disc engine, installed at the office of *The Times*.[61]

Agitation for patent law reform grew as the Great Exhibition approached, and Farey joined the committee set up by the Society of Arts to make representation to the government.[62] Notwithstanding that he had given the greater part of the evidence to the Committee on Patent Law Reform of 1829, and had given evidence to the Lords Committee in 1835, he did not appear before the Parliamentary Committee to investigate the affairs of the Signet and Privy Seal Offices of 1849, which investigated the mechanics of obtaining a patent. It seems he did not take a very active part in the deliberations of the Society of Arts Committee either.

Misfortune struck him again in 1850, for on 15 November his wife Elizabeth died, and was buried at Highgate Cemetery. In December of that year Farey arranged for a fair copy of his will to be made, but made no change to the wording. Since Edward was now the sole beneficiary it might have been more sensible to have rewritten it, but it was left alone.

Farey was appointed a juror to the machinery section of the Great Exhibition in June 1851, and was active in his duties. He noted particulars of some models of early steam engines he had seen at the exhibition, with their class numbers, in the proof copy of the second volume of his *Treatise*.[63]

In July 1851 Farey visited Sevenoaks, Kent, and on 17 July he died there of heart failure. He was buried with Elizabeth at Highgate, where their grave may still be seen in the unconsecrated section of the closed part of the Cemetery.[64]

No contemporary obituary of Farey has been found, and the only comment about his character appears in a long memoir printed in *Proceedings of the Institution of Civil Engineers*, which concluded by describing his as a 'good and upright man ... beloved ... by all who knew him'.[65] This memoir dwells largely on his early life and his writings, and largely ignores his later consultancy work.

Farey's will was proved on 3 December 1851 with a gross value of £11,571 4s 11d. Elizabeth was dead and his other executor, Joshua Field, of Maudslay, Son & Field, declined to act, so Edward was both sole beneficiary and executor.[66] It is not known whether the listing of the various bequests which Farey had willed was ever made, but a recent legal opinion is that Edward would have been required to make a list in order that the will could be proved.[67]

The 1886 edition of the *Catalogue of the Patent Office Library* listed a manuscript catalogue of Farey's library, but this is now lost.[68]

Edward was described in the 1851 census as an assistant to his father, but he seems to have given up engineering after his father's death and by 1872 he moved to 12 Greville Place, Kilburn, London, which was to remain his home for the rest of his life. The Post Office Directories from 1853 list Edward only on the Court pages, with the rank of Esquire, and this indicates that he did not work, being of independent means. He is said to have once been jilted, after going to the expense of laying a special dance floor for his young lady, and in his later life he seems to have been a rather sad, reclusive figure.[69] A photograph taken when he was aged about 35 shows extensive retouching in the angle of the jaw, probably to remove a scrofula blemish he may have carried since his birth.[70]

Since Edward was evidently not working as an engineer, he would not have had much use for his father's collection, and he appears to have disposed of it in some way, notwithstanding that John Farey's will directed that it was to be kept together. The only sale of an engineer's library so far traced shortly after Farey's death was that held on 2 June 1852 by Puttick & Simpson, the London book auctioneers, who sold 'The Library of an Eminent Engineer'. The four hundred or so lots are a very eclectic collection indeed, for not only were there books about mostly civil engineering, but also books about heraldry and genealogy, literature, book collecting and theology – just the sort of topics Farey is known to have been interested in.[71] It is possible that the collection was the residue of Farey's library, the section relating to his patents practice having been sold privately beforehand. The question remains, however, of what then happened to his library and working papers.[72]

Farey was clearly a product of his age and was typical of that utilitarian-cum-radical style of Englishman who did much to foster social progress. Nothing is known of his religious beliefs, but there is a strong thread of Unitarianism in a number of his relationships, among his friends and among his professional associates. He was a close friend of John Taylor (1779–1863), the mining engineer, and his brother Philip (1786–1870), the mechanical engineer. As a young man Farey had written and drawn for publications edited by the Unitarians Abraham Rees (1743–1825) and Jeremiah Joyce (1763–1816). Farey took as a mature pupil William Bridges Adams (1779–1872), who was the husband of Sarah Flower Adams (1805–48), the Unitarian hymn writer. Bridges Adams himself was the author of polemical essays on social questions in Unitarian periodicals.[73]

There is a thread of radicalism in Farey's circle. His friend and client Joseph Chessborough Dyer (1780–1871) was an active radical and was greatly involved in the Anti-Corn Law League. The links between the Dyer and Farey families were of many years' standing. John Jr had drawn up many of Dyer's patents, and as a young man Henry worked for a while in Dyer's Manchester factory. Dyer was an American who moved to Britain in 1811 and set himself up as a patents farmer, importing American inventions and improving them before obtaining a British patent for the inventor and then retaining a half-share in any profits. He brought over, among others, Amos Whittemore's wire-card making machine (no. 3498) and Danforth's flyer (no. 5909). He was one of the founders of the *North American Review* and also the *Manchester Guardian*.[74]

If it is assumed that a person's choice of friends is an indication of his or her own philosophical and political leanings, it is highly probable that Farey was, at the very least, a nonconformist in the widest sense, and may well have been a Unitarian as well, in spirit if not in practice.

Farey's personality is difficult to assess, for no private papers have yet been discovered. As a young man he evidently had boundless self-confidence and energy, but this seems to have been eroded as he got older, perhaps as his health deteriorated. He was avid for facts and meticulous in publishing detail, perhaps to a fault. He was obviously in the same mould as his father, interested in everything and not afraid of stating his opinions.

He was highly intelligent and had a natural bent to scholarship in the widest sense. This is clear from his writings; from his scientific approach in collecting data from which design formulae could be established; from his ability to write and speak both French and Latin; and, in his historical work, from his recourse to both classical and modern writers.

Farey was a professional backroom-worker, whose high eminence was known largely to parliamentarians, publishers, manufacturers, engineers, patentees and lawyers, but not to the public at large. He offered a highly specialized technical service of immense value to the progress of industrialization in Britain. He deserves far more honour for this than he received from his peers and from posterity.

Few of his contemporaries appear to have approached him in the breadth of his historical knowledge; the only other who springs to mind is John Percy (1816–89), whose books on metallurgy, published from 1861 onwards, are very similar to Farey's in their comprehensiveness.[75]

It is curious that such a talented man as John Farey should have been so neglected since his death. Apart from his obituary in the *Proceedings of the Institution of Civil Engineers*, published in 1852, and his *Dictionary of National Biography* entry, no other biography seems to have been published about him.

In many ways Farey resembled Isambard Kingdom Brunel: they were encouraged by their fathers from an early age and both moved in the highest scientific and intellectual circles from their youth; both were born in an age when it was not necessary to have an extensive academic training before starting a career in engineering; both were ignored by Samuel Smiles, though his neglect of them is not hard to understand, since neither followed the rags to riches progression of the majority of the people of whom he wrote.

Here, however, the similarities between Brunel and Farey end, for Brunel's surviving work – his railways and bridges – have ensured a continual fascination for technical historians. In contrast, Farey's work was largely unknown to the public at large, working as he did so often as midwife to other men's ideas. Perhaps the overwhelming reason for this disparity of interest between the two men is that no contemporary biography of Farey ever appeared to inspire later workers. This conclusion is reinforced by posterity's view of Farey's contemporaries Faraday and Babbage. Both men were born in the same year as Farey, 1791. Both received great public recognition during their lifetimes, and both, in consequence, were accorded much attention in 1991, the bicentenary of the year of their birth.[76]

There is the question of how much Farey's own modesty affected his later reputation. The bulk of his writings were published anonymously or pseudonymously, and there is a vast amount of his work locked up in the texts of his patent specifications, which remain an untapped source. According to Edward Farey's brief biography, his father once wrote extensively on theology, but nothing has so far been traced to him. Farey's own application for election to the Royal Society is further evidence for his inability to advance his own case, concentrating as it does on his early writing and drawing work, and missing out completely any detail of his work as a consulting engineer. However, there may be justification for his low profile in this instance, since so much of his patent consultancy and perhaps other engineering work would have been confidential, in contrast to a scientist whose goal is to publish every discovery he makes, in order to gain priority over fellow workers.[77] It is this anonymity that has caused John Farey to be forgotten.

That Farey has for so long been neglected is not a true reflection of his importance, but is more a reflection of the view technical historians have of their field of study. On the one hand there is the highly questionable assumption that someone for whom there is not a comprehensive body of records cannot be written about. This is nonsense for more recent times, since there is a wealth of records, albeit fragmentary, available which can be used to construct much of the skeleton of a person's life. On the other hand there is the undoubted fact that the greatest amount of interest on the part of technical historians today seems to be directed to civil engineering remains such as railways and canals, rather than the industrial processes and machinery of the industrial revolution. This concentration on civil engineering and in particular railways is a continuation of the railway mania which swept Britain in the 1840s, and which placed transportation on such a high pedestal. But improvements in transportation technology are the end of the chain on industrial technology. Without the improvements in production technology, which generated more wealth for the nation, there would not have been enough capital to fund the improvements in transportation. The people who made the improvements in production technology possible deserve just as much investigation and acclaim as their colleagues who specialized in civil engineering projects.

Biographical studies attract minimal interest: it is extraordinary that a society which has been so heavily based on technology and industry for the past two centuries can still show so little regard for the men who actually made it happen.

The diffusion of Farey's writings and drawings throughout the engineering and manufacturing community may be presumed to have had an immense impact in spreading knowledge of new machines and processes, which in turn must have had some influence on Britain's industrial and economic development. His detailed knowledge of technical history would have prevented many inventors from having their patents rejected on points of law, and so enabled new inventions to be introduced more rapidly.

Farey was a pioneer of the systematic study of technical history, as he described in a letter to Macvey Napier, written in 1814:

> You will readily believe this from the nature of my profession as an engineer, in which I have devoted my time to a field never before explored to any great extent. This is research into the history, practice and theory of mechanical inventions for the assistance of inventors who wish to procure patents. I have in the past 4 or 5 years been principally employed by these different persons as to give advice on the probability of success and into the history as to determine whether a patent-right can be secured to them as the first inventors.[78]

J.C. Dyer was Farey's contemporary and knew him well; it is fitting to allow him the last word:

> Mr Farey was for some time an able assistant to Mr Heathcoat in his plans for improving lace-making machines, and also in aiding Mr William Eaton in constructing his self-acting mule for cotton spinning. Having elsewhere[79] treated of these several inventions of Mr Heathcoat and Mr Eaton, I notice them here merely to show the varied and superior talents of Mr Farey and their due appreciation by the most eminent of his contemporaries engaged in mechanical science, so that the name of John Farey should ever stand high among those then contributing to this branch of knowledge. ... The early training and precocious mental powers of Mr Farey may have checked the development of his physical constitution, which was but feeble, and led to his early death, in the midst of his useful labours at about sixty, I believe. Those whose labours fructify for the general good, long after their death, should have their names transmitted among succeeded generations; but the fashion of the time is to bow down before some idol of the day and see him replaced tomorrow by a new one.[80]

APPENDIX: THE FAREY FAMILY

John Farey Jr was born on 20 March 1791 at Lambeth, London, the eldest child of John Farey Sr (1766–1826) and Sophia Hubert (1700–1830).[81] His father's family originated from Bedfordshire farming stock, but from 1763 his grandfather, also John Farey, was working at Westminster, London, as a carpenter and builder.[82] His mother's relatives included a number of London architects and surveyors; her brother Samuel was a surveyor to the Sun Fire Insurance Company at the beginning of the nineteenth century, and later worked as an architect, while her sister, Mary, was the wife of Thomas Cundy (1765–1825), the architect and builder. Their son James (1793–1825) and grandson Samuel (1816–67) were both well-known masons and sculptors.[83] The only brother of John Farey Sr, Benjamin (b. 1776), was a successful road surveyor practising around London, and of his family, two daughters married into the Colman family, of mustard fame, while his great-grandson, Cyril Arthur Farey (1888–1954), made a name for himself as an architect and draughtsman during the first half of the twentieth century.[84]

John Farey Sr was born at Woburn, Bedfordshire, and after elementary education there was sent to Pullman's Academy at Halifax, Yorkshire, where he made the acquaintance of John Smeaton, the eminent engineer, who lived nearby at Austhorpe. At Pullman's Academy, the master 'gave him gratuitous instruction in mathematics and philosophy',[85] and he was also given teaching in drawing and surveying. As a youth, John Farey Sr was evidently much interested by the steam engine, since John Jr noted that he had his father's drawings of the engine at Seacroft Ironworks, sketched in 1782, and of the engine at Whitbread's Brewery, London, sketched in 1785.[86] Smeaton may well have had an influence on John Farey Sr in his choice of career, since he gave advice to Farey's great friend Bryan Donkin Sr (1768–1855), later to become a famous engineer, about obtaining an engineering apprenticeship.[87]

It is now known when he moved to London after finishing his education, but it must have occurred at the end of the 1780s. He probably used his family connections to obtain work as a surveyor. Nothing has been found to show how he came to meet Sophia Hubert, but he was well enough established in his profession to marry her at St Pancras Church, London, on 19 May 1790. John and Sophia were both keen singers, with John singing tenor and his wife soprano. They performed with the Surrey Chapel Society and the Cecilian Society, and took part in oratorios at Drury Lane Theatre. They also sang with the choir of the London Choral Fund, a group of amateur singers that gave concerts in aid of destitute singers and their dependants, and in August 1791 John was appointed the Fund's secretary and librarian.[88]

In 1792 John Farey Sr was appointed land steward to the Woburn estate of Francis, the 5th Duke of Bedford. The Duke was Whig in politics and was an active supporter in the House of Lords of Charles James Fox. He was a bachelor and, according to Arthur Young, godless.[89] He was one of the founding members of the Board of Agriculture. John Farey Sr was thus drawn into the very centre of the movement for agricultural improvement then taking place in England. He was concerned with developing the model farm the Duke constructed at Woburn, was involved with the renowned Woburn sheep-shearing trials and also with the Smithfield Club, which the Duke helped found in 1798.[90] Farey wrote articles for the farming press and was involved in forestry experiments sponsored by the Society of Arts in 1794.[91]

Farey and his family lived at Pottsgrove, the family home, about three miles from Woburn Abbey, and while there Sophia gave birth to five more children: Sophia (1793–1843), William (1795–1836), Joseph (1796–1829), Marianne (1799–1883) and Henry (1800–58). John Jr commenced his education by attending the local dame's school there, although as will be seen later it is clear that he received much informal education from his father, who took him, as a very young child, to attend lectures given by itinerant teachers, perhaps at Leighton Buzzard.[92]

In October 1801 John Farey Sr became acquainted with William Smith (1769–1839), the Father of British Geology, whose theories of stratification he later publicized, and it was while on a field trip with Smith, in March

1802, that he heard of the sudden death of the Duke of Bedford at the age of 37. Farey was forced to leave his post immediately, since there was no place for him in the employment of the 6th Duke. The reasons for this dismissal are not known, but the event was noted in several letters discussing the family history in the possession of the American branch of the family, and so have come down through word of mouth.

Now must have commenced a very worrying time for John and Sophia Farey and their family, for but for the death of their employer they could reasonably have looked forward to living at Woburn for many years. The family moved to London, and at first thought about emigrating. Then, in November 1802, John Sr wrote to Arthur Young that he was contemplating taking a farm somewhere.[93] This plan came to nothing, and John Sr set himself up as a mineralogical surveyor and author. The family lived at 12 Crown Street, Westminster (the site today of the Home and Foreign Offices), and this was to remain their home until March 1816. Sophia later gave birth to more children: Ann (1803–69) and two who died in infancy, Francis (1807) and Frances (1808–9), In 1805 John Sr was appointed secretary of the Smithfield Club in succession to Arthur Young. This was originally formed in 1789 as the Smithfield Cattle and Sheep Society with the object of improving the quality and lowering the price of meat sold at Smithfield Market.

The first decade of the nineteenth century was an important one for the world of learning, and in particular for the growth of scientific institutions and investigations into the scientific background of technology. John Farey Sr was in great demand as a consultant by landowners wishing to exploit their property, and was in contact with many of the scientific luminaries of the time.

He was a prolific writer. The Royal Society's *Catalogue of Scientific Papers* listed around 60 papers by him, but recent work by Hugh Torrens has more than tripled this.[94] For Rees's *Cyclopaedia* alone he contributed articles on canals, geology, mineralogy, surveying and a number on the scientific and mathematical basis of sound. His other writings were on all manner of topics, ranging from horticulture, geology, meteorology, metrology, currency decimalization, music and mathematics, to peace and other topics then of interest. In 1807, through the influence of Sir Joseph Banks, he was given the job of writing for the Board of Agriculture an agricultural survey of Derbyshire, which was published as *General View of the Agriculture of Derbyshire* (3 volumes, 1811–17).[95] It is for this book and for his work on geology that he is remembered by historians of geology today,[96] but the family name is assured a more widespread and permanent memorial by the Farey numbers, used by mathematicians, whose properties he observed as a consequence of his interest in numerical series, stimulated by his investigation into the mathematics of sound.[97]

The Farey family operated a private business, servicing the various clients who wished to make use of their skills. John Sr concentrated on his surveying and on writing for the scientific and popular press. John Jr was greatly helped by his younger brothers and sisters. Joseph and Henry were engineers assisting inventors by preparing patent specifications and building prototype models, as well as making drawings and writing articles for publishers. William was a land surveyor and probably helped his father for at least part of his time. The girls, Sophia, Marianne and Ann, helped by copying and colouring technical drawings as well as making drawings in their own right for fashion magazines.

How it was organized was briefly described by John Farey Sr in a letter to Davies Giddy written on 26 March 1816:

> Since I had the honour of writing to you on the 21st, I have made some progress in removing to a house here [37 Howland Street, Fitzroy Square] more commensurate with the number and growing size of my family and to my eldest son's business, as an engineer and writer etc. ... (in which he employs 2 and sometimes 4 of my other children, under my own or his mother's eye), the better to accommodate our business.[98]

William took over from his father as Secretary of the Smithfield Club on the latter's death in 1826, and retained the post until his own death ten years later.[99] William left a young family, and his only son, Barnard William Farey (1827–88), later became a well-known engineer.

Considerable obscurity surrounds the life of Henry Farey. In the 1820s he was in France working with his Gengembre connections, who by that time had established themselves as engineers in Paris. Later he became manager of a machine manufactory set up at Gamaches (Somme) by Joseph Chessborough Dyer and his sons. The Gamaches concern failed and in later years Dyer's sons publicly blamed Henry's gross mismanagement for the loss of £120,000. Dyer himself, in his 'Memoir', was extremely generous in his praise of John Farey Jr and of his abilities, and did not mention the Gamaches concern at all.[100] Henry emigrated to the United States sometime in the late 1830s and by 1850 was settled with a family at Maquon, Illinois, where he was working as an engineer. He died in 1858 at Knoxville, Illinois.[101]

Sophia Farey never married and she died at Twickenham in 1843.[102] Ann spent most of her adult life as the favourite maiden aunt of the children of her sister Marianne Gengembre, first in France and then in the United States, where the Gengembres had emigrated in 1849.[103] Colomb Gengembre had been active in French radical politics, and the revolution of 1848 may have caused him to flee the country with his family.[104] His two sons helped in the design of river steamboats for Sherman's Mississippi campaign in the American Civil War. Marianne Gengembre moved to New York to be with her children (who by that time had added the name Hubert to their French one, from the maiden name of Sophia, the wife of John Farey Sr) and she died there in February 1883.[105]

Edward Farey was groomed to become assistant to his father, and in 1842 was an architecture student at University College, London.[106] His cousin Barnard William (1827–88) was sent in 1836 – at the age of 9 – to his uncle Colomb, who by this time had succeeded his father as director of the French naval engine manufactory at Indret on the Loire. In 1843 Barnard William, then aged 16, returned to England, and spent some time at the Fairfield Works, Bow, London, owned by William Bridges Adams, where locomotives were made. He later joined Donkins, the engineers of Bermondsey, where he was ultimately appointed a partner in the company and was responsible for a number of inventions. He died, married but without issue, in 1888.[107]

The Gengembre branch of the family evidently kept in touch with Edward, and in the 1870s some of them travelled to England and called on him when they were in London.[108] In 1888, Edward was executor to the will of his cousin Barnard William Farey, who died in that year.

The biographies of John Farey Sr and John Farey Jr were published in the *Dictionary of National Biography* in 1889. That of John Jr was written by G.C. Boase and was based on the one published by the Institution of Civil Engineers in 1852. Boase added the erroneous date of 1850 for the year of the fire, and also noted Farey's place of death; neither of these details had appeared in the earlier biography.

In 1898, through a chance reading of the Farey name in a local newspaper, the Gengembre branch of the family made contact with the descendants of Henry Farey, who by this time had moved across America to settle in California, where the Gengembres had moved a few years previously.[109] The two American branches of the family kept in touch in succeeding years, and from the surviving correspondence it is clear that they were well aware of their illustrious ancestors. Kate Hubert (granddaughter of Marianne Farey) and Mary Farey (granddaughter of Henry Farey) compiled a tree of the Farey and Gengembre families (Figure 5), and it is plain they had access to more historical material than now survives. Henry Farey's diaries – at least, such of them as remained – were in the possession of the Farey branch. The cousins exchanged relics and items became lost in the process. For example, Kate Hubert gave Henry's diary for 1818 to an unnamed cousin on the Gengembre side,[110] while as recently as 1941, when an old lady in her 89th year, she returned to Mary Farey Henry's diaries for 1825 and 1828, now also lost.[111]

Edward Farey kept in contact with Caroline Holroyd, his stepmother's unmarried niece, and he was friendly with William James Russell, FRS (1830–1909). Russell was an eminent chemist, but just how he and Edward became friends is not yet known.[112] He was also friendly with his lawyer, Edward Pearson, whom he used to visit with some frequency. A surviving daughter remembered how he used to go through the house winding all the clocks but never once speaking to any of the children, and how he used to play billiards with her father. The Pearson children used to remark on how unfair Edward was in making his coachman stay on the box of the coach in the street in all weathers while waiting for his master.[113]

Whatever the fate of his father's library, Edward certainly retained some items. He died, unmarried, on 15 October 1913 at the age of 88, and willed to the Institution of Civil Engineers 'the printed but unpublished volume of

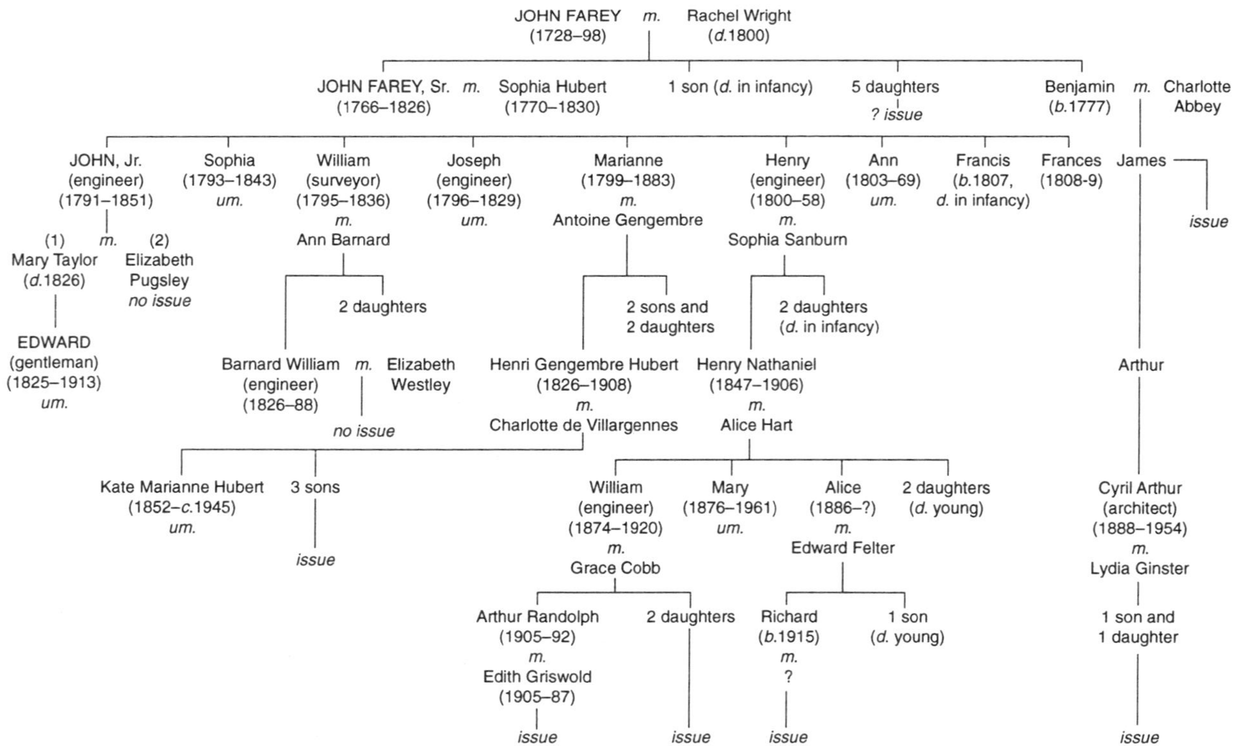

Figure 5 The Farey genealogy. This is not a complete family tree, as it lacks much detail of the family of Benjamin and the present-day American branch. Based on data collected by Kate Hubert and Mary Farey.

the work on the steam engine by my father John Farey'. For various reasons it never got there, but appeared at an auction a little later, to be purchased by the Patent Office Library. Edward willed to the Royal Society the six volumes of manuscript drawings known as Smeaton's *Designs* and his father's annotated copy of Smeaton's *Eddystone* (1814). All his other property (apart from some minor personal effects and his clothing) was to be sold, and the proceeds split among beneficiaries. Kate Hubert was to receive £500. The descendants of Henry Farey were ignored, though he must have known of their existence, assuming Kate Hubert had mentioned them in her letters. His residual estate was split between Caroline Holroyd, the children of his friend William Russell, FRS, and University College Hospital, London. His servants received annuities. The will makes no mention of any other books or manuscripts. The estate was assessed for probate at a value of more than £40,000.[114] The Royal Society correctly received its bequest, but placed it in store, where it remained until the late 1930s, by which time all knowledge of the history of the drawings had been lost.[115]

In addition to her money, Kate Hubert was sent a number of other effects, including daguerreotypes of John Farey Jr and Ann Farey, John and Mary Taylor's marriage certificate, a Soho-pattern slide-rule and the baptism certificate of Edward. It is also probable that the two volumes of sketches of machinery made by John Farey in 1804–10 and some of the proofs of Farey's drawings for the unpublished volume of his *Treatise*, all now in the United States, came there by this means.[116]

The set of drawing instruments presented to John Farey in 1844 by Robert Stephenson was sent by the lawyer to Alexander Scott, FRS, and his wife, who was the daughter of William Russell. These instruments passed to the Royal Society of Edinburgh on Alexander Scott's death in 1947, and were sold to the Royal Scottish Museum in 1990, together with a brief biographical outline of John Farey's life by Edward – reproduced at the beginning of this paper.

The sale of Edward's house and property probably occurred early in 1914. The early records of Edward's solicitor were destroyed in an office move during the 1970s, just before the author's research into Farey began, so there is no longer any way of readily discovering the details of how Edward Farey's property was dispersed and whether it included any more of his father's relics and papers.[117]

Acknowledgements

I am grateful to many people for help; in particular I owe an immense debt to the late Arthur and the late Edith Farey for generously providing copies of documents in their possession. I am likewise in the debt of Messrs Longmans for giving permission to make use of their records, and to the Trustees of the British Library for permitting copies to be made of documents in the Department of Manuscripts. The Massachusetts State Historical Society is also thanked for permission to make quotations from the unpublished 'Memoir' of Joseph Chessborough Dyer.

This work would never have been done without the help of many librarians, in particular Vicky Tattle and her colleague John Hewish of the Science Reference Library; Peter Jones, late of the Department of Manuscripts of the British Library; L.R. Day, late of the Science Museum Library; M.M. Chrimes of the library of the Institution of Civil Engineers; Dr J.A. Edwardes, Archivist to the University of Reading and custodian of the Longmans archive; and Mrs J. Percival of the library of University College London. I owe a particular debt to my friend David Bromwich, Somerset Studies Librarian, Taunton Library, Somerset. In addition I salute the ever-efficient staffs of the Public Record Office and of the Principal Probate Registry, London.

I acknowledge the help given by fellow members of the Newcomen Society. Professor Alec Skempton, James Harrison, the late John Hills and the late George Watkins all read various early drafts and made valuable comments. The late Charles Lee provided the clue which led to the American branch of the Farey family.

Hugh Torrens has supported my work continuously by making freely available his immense knowledge of the scientific world at the beginning of the nineteenth century, and for sharing his information about the life of John Farey Sr. We have in hand a study of the Farey family as a whole.

Finally, none of this work would have been possible without the support of my wife Jane and my sons Matthew and Dominic.

Notes and References

1. John Farey, *Treatise on the Steam Engine*, Vol. 2 (Newton Abbot, 1971), p. 7, note a; A.E. Musson and Eric Robinson, *Science and Technology in the Industrial Revolution* (Manchester, 1969), ch. 3.

2. Anon., 'Memoir of John Farey, jr.', *Proceedings of the Institution of Civil Engineers*, 1851–2, XI: 100–2.

3. James Harrison, 'Some patent practitioners associated with the Society of Arts', in *Journal of the Royal Society of Arts*, 1982, 591–2; William Nicholson, *Journal of Natural Philosophy and the Arts*, 1804, 7: 67–77.

4. However, Anon., *A Slight Sketch of Dr Andrew Ure* (London, 1874), p. 6 mentions that 'Mr Farey' was a pupil of Ure at the Andersonian Institution, Glasgow. Ure taught there between 1804 and 1809, and it is thus possible for John Farey to have studied under him. But this book then goes on to say that others of Ure's students were John Beaumont Neilson, who in fact did not enter the Institution until 1817, after Ure's departure (Samuel Smiles, *Industrial Biography*, ch. IX), and William Fairbairn, who made no mention in his *Autobiography* of having studied there. So more evidence will be needed before this claim can be substantiated.

5. In the possession of the Farey family of California; microfilm copy in the Science Museum Library, London.

6. A.P. Woolrich, 'John Farey, jr. and Rees' *Cyclopaedia*', *Industrial Archaeology Review* (forthcoming).

7. A.P. Woolrich, 'The diaries of Henry Farey, 1816–1820' (forthcoming).

8. Sir John Rennie, Presidential address to the Institution of Civil Engineers, 1846, printed in the Institution's *Proceedings*, 1847, 7: 62.

9. John Farey Sr, 'On the alleged plagiarisms of Frenchmen . . .', in *Philosophical Magazine*, 1820, 55: 321–3; John Farey, *Treatise on the Steam Engine*, Vol. 1 (London, 1827), p. 653, note b.

10. Henry Farey's diary (see note 7), various entries for September 1816. John Heathcoat had six French patents between 1820 and 1828, and Farey and the sons of William Eaton took out a joint patent for the mule jenny in November 1823. Information from the Institut National de la Propriété Industrielle, Paris.

11. The family has been investigated by Mr Ian Frazer, who is studying French radical politics in the nineteenth century.

12. House of Lords Library, MS of evidence given to the Select Committee on Patent Law, August–September 1835, pp. 14–15.

13. He engaged the engineer John McNaught of Glasgow to manufacture them in the early 1820s. Ronald M. Birse, 'Life of William McNaught (1813–81)' in *The Dictionary of National Biography: Missing Persons* (Oxford, 1993) 436–7. William McNaught was the inventor of 'McNaughting' of low-pressure Watt engines by the addition of a small-bore high-pressure cylinder. The first printed reference to Watt's indicator appears to be a letter in *Quarterly Journal of Science*, 1822. H.W. Dickinson in his *Short History of the Steam Engine* (Cambridge, 1938), p. 85, wrongly stated that Farey saw the device in 1826, and this mistake has been repeated by later writers, such as D.S.L. Cardwell, *From Watt to Clausius* (London, 1971), 80.

14. Henry Farey's diaries, various entries for June–September 1819.

15. *Report of the Committee on the Cheapest, Most Expeditious and Most Efficient Mode of Providing Small Arms for Her Majesty's Service* (London, 1854), 172–3.

16. British Library, Additional Manuscripts, 34614, f. 335.

17. Benett Woodcroft, *Titles of Patents of Inventions Chronologically Arranged*, 2nd ed. (London, 1857).

18. Anon., 'Notices respecting New Books: The Cyclopaedia ...', in *Philosophical Magazine*, 1820, 56: 224.

19. A.P. Woolrich, 'John Farey's *Treatise on the Steam Engine*, 1827', *History of Technology* (forthcoming).

20. University College Library, London, Archives of the Society for the Diffusion of Useful Knowledge.

21. Farey, *Treatise on the Steam Engine*, Vol. 1 *op. cit.* (9), v–vi.

22. James Sowerby, *Mineral Conchology of Great Britain*, Vol. 3, (London, 1822) 188, where is printed a letter from John Farey Sr which incidentally mentions dropping 'Sr' and 'Jr' from their writings.

23. Marriage registration certificate, in the possession of the Farey family, California.

24. W.G. Rimmer, *Marshalls of Leeds, Flax Spinners, 1788–1886* (Cambridge, 1960), 150ff.

25. Edward's baptism certificate, in the possession of the Farey family, California.

26. MS notes on family tree, in the possession of the Farey family, California.

27. Joseph Farey became paralysed at the time he was taking out his patent for an improved oil lamp (no. 5214, granted 25 July 1825). John Farey cited his brother's difficulty in not being able to attend personally to enrol his specification in evidence he gave to the Parliamentary Committee on Patent Law. *Report of the Parliamentary Select Committee on the Law Relating to Patents for Inventions*, 1829, CMND 332, p. 37.

28. Rupert Gunnis, *Dictionary of British Sculptors* (London, 1961), 118.

29. Professor Sir Alfred Pugsley, 'The Pugsley Family', in *Devon and Cornwall Notes and Queries*, 1973, 32: 203–10. For the Holroyd family see Venn's *Alumni Cantab, 1752–1900* (Cambridge, 1940–54), p. 426; *Australian Dictionary of Biography* (Melbourne, 1972), Vol. 4, 412–13; *Burke's Colonial Gentry* (London 1891), 30–1.

30. The death certificate of Joseph Farey records his dying of paralysis (information from Ian Frazer). The story of the barber's infected razor occurs in several of the letters written by his great-niece, Kate Hubert, early in the twentieth century, and now in the possession of the Farey family, of California.

31. University College Library, London, Archives of the Society for the Diffusion of Useful Knowledge.

32. *Mechanics Magazine*, 1830, 13: 308–12, 377–89. The case continued until 1831, and was reported in *Mechanics Magazine*, 1831, Vol. 15, with more of Farey's evidence at pp. 418–25 and 435–39. Farey again came in for adverse comment from the editor.

33. University College Library, London, Archives of the Society for the Diffusion of Useful Knowledge.

34. *Observer*, 22 December 1844; Nottingham University, Fisher MSS (addresses on various bills for patents work). In Farey's time contemporary documents (including the draft manuscript of Farey's will) spelled 'Guilford' as 'Guildford'.

35. *Proceedings of the Institution of Civil Engineers*, 1851, XII, 101–2.

36. The book, *A Practical Treatise on the Law of Patents* (London, 1830), is not cited in the 1955 edition of the *Catalogue of the British Museum Library*.

37. Institution of Civil Engineers Library: Clarke MSS [unpublished history of the

Institution], pp. 21ff. Farey's detailed schemes and accompanying correspondence also survive in the library.

38. Report of the inquest in *Civil Engineer and Architect's Journal*, 1838, 1: 284–7.

39. Science Museum Library, London: Field MSS, report on round and square fireboxes in locomotives by John Farey and Joshua Field to the Directors of the Liverpool and Manchester Railway, 1831/32.

40. John Farey, *Erard's New Patent Action Grand Pianoforte* (London, 1836) (Patent Office Library Catalogue no. 2587).

41. Patented Oct. 1833 (no. 6484).

42. William Felkin, *History of Machine-Wrought Hosiery and Lace Manufactures* (London, 1867), 324–5.

43. *Mechanics Magazine* 1836, 24: 379–80.

44. H.I. Dutton, *The Patent System and Inventive Activity during the Industrial Revolution, 1750–1852* (Manchester, 1984), 49. For some reason Dutton claims that Farey was a Scot!

45. Nottingham University, Fisher MSS, Fi A 15/10.

46. Brian Bowers, *Sir Charles Wheatstone, FRS, 1802–1875* (London, 1970), p. 111.

47. Geoffrey Hubbard, *Cooke & Wheatstone and the Invention of the Electric Telegraph* (London, 1965), 58.

48. The original daguerreotype has been presented by the Farey family to the National Portrait Gallery.

49. Little is known about how Farey conducted these tests. See note 61 for one of his reports. His nephew, Barnard William, was later to become involved in engine tests with his partner, Bryan Donkin Jr (1835–1902), in the 1860s and 1870s. They built a special steam engine which allowed them to measure the temperature of the steam leaving the boiler and the temperature of the condensate leaving the engine. The difference in temperature was a measure of the amount of heat converted into energy by the engine. The merits of an engine were thus expressed in terms of the ratio which the whole heat converted into work bore to the heat in the steam supplied by the boiler. Anon., Obituary of Bryan Donkin, *The Engineer*, 14 March 1902, p. 258. This paper, incidentally, confuses John Farey Jr with Barnard William. There was a succession of several Bryan Donkins.

50. Public Record Office (PRO): will of John Farey, proved PCC December 1851.

51. Jacob Samuda, 'The Atmospheric Railway', in *Proceedings of the Institution of Civil Engineers*, 1844, 3: 277–82, 283.

52. *Mechanics Magazine*, 1844, 41: 447.

53. *Ibid.*, p. 311.

54. *The Times*, 31 December 1844.

55. *Ibid.*, 30 December 1844.

56. *Observer*, 29 December 1844.

57. Royal Society Archives, London: Farey's 'Certificate of a Candidate for Election', dated 22 February 1847.

58. Bryan Donkin [engineer]; Ted Pollock [Sir Edward Pollock, lawyer]; John Rennie [engineer]; Joseph Miller [not in *Dictionary of National Biography*]; William Shadwell Mylne [engineer and architect]; John Taylor [engineer]; Joshua Field [engineer]; William Brockedon [painter, author and inventor]; Andrew Ure [technical writer]; W. Ayrton [musical writer and critic]; W. Cubitt [engineer]; Phillip Hardwick [architect]; William Buckland [geologist]; J.M. Rendel [engineer]; T.S. Brandreth [mathematician, classicist and barrister]; Geo. Lowe [not in *Dictionary of National Biography*]; plus one illegible signature.

59. *Proceedings of the Institution of Civil Engineers*, 1849, 8: 150.

60. Isaac Babbitt, *Reports of Babbitt's Patent Lined Bearings* (London, 1848). The composition of Babbitt metals was varied according to the duty the bearing was to perform. Some mixes were largely tin with small amounts of antimony and copper. Others were largely copper with a small amount of zinc. *Kempe's Engineer's Year-book* (London, 1913), 613.

61. Details of this engine were published in *Practical Mechanics Journal*, 1849, 2: 98ff. Farey's report on its performance was published in *Engineering*, 15 February 1884, whence it had been sent by Bryan Donkin Jr, perhaps via Barnard William Farey, who was a Director of the Donkin concern.

62. H.I. Dutton, *The Patent System and Inventive Activity* (Manchester, 1984), 58.

63. John Farey, *Treatise on the Steam Engine*, Vol. 2 (Newton Abbot, 1971), 7, 8.

64. Camden Library, London, Records of Highgate Cemetery. Grave 3864, square 31.
65. *Proceedings of the Institution of Civil Engineers*, 1852, 11: 102.
66. PRO: records of PCC, Admon attached to PCC copy of Farey's will.
67. Information from Mr David Tew, Hon. Solicitor to the Newcomen Society.
68. 'John Farey, catalogue of his library, MS', Patent Office Library (ref. no. 12677).
69. Information from Mr Michael Pearson. His grandfather had been Edward Farey's solicitor and was a personal friend.
70. A photograph taken in London in 1861, now in the possession of the Farey family, California. When Edward was a baby he suffered badly from scrofula, which is a tuberculosis of the bones and lymphatic glands, and can cause disfigurement. His father spent a great deal of effort in trying to find a cure both in England and France. Eventually he used a remedy made by a Mrs Knight which worked well, and Farey cited his experience with this patent medicine to the 1829 inquiry on patent law. *Report of the Parliamentary Select Committee on the Law Relating to Patents for Inventions*, 1829, CMND 332, p. 135.
71. British Library, printed catalogue of the Puttick & Simpson sale.
72. A.P. Woolrich, 'John Farey and the Smeaton Manuscripts', in *History of Technology*, 1985, 10: 181–216.
73. For Taylor, see Roger Burt, *John Taylor* (Hartington, Derbyshire, 1977); for Bridges Adams and his wife, Rees and Joyce, see the *Dictionary of National Biography.*
74. For Dyer, see the *Dictionary of National Biography* and Shirley McKenna, 'Joseph Chessborough Dyer', in *Memoirs and Proceedings of the Manchester Literary and Philosophical Society*, 1974–5, 112: 104–11.
75. John Percy, MD, FRS, *Fuels; Fireclay; Copper; Zinc; Brass* 1861; *Iron and Steel* 1864; *Lead* 1870; *Gold and Silver* 1880. These have been reprinted by De Archaeologische Pers (Eindhoven, Netherlands), from 1985.
76. Babbage-Faraday Bicentenary Conference, Cambridge, 5–7 July 1991, *Transactions of the Newcomen Society*, 1990–1, 62: 143–56.
77. I am grateful to Dr Hugh Torrens for this insight.
78. British Library, Additional Manuscripts, 34611, fol. 43.
79. *Memoirs and Proceedings of the Manchester Literary and Philosophical Society* contain a series of papers by Dyer on mechanical topics:

'Notes on the introduction of steam navigation', read 10 February 1863 and published in Vol. II, ser. III, 1864, pp. 284–95.

'Notes on the origin of several mechanical inventions, and their subsequent application to different purposes'.

Part 1: read 17 October 1865 and published in Vol. III, ser. III, 1865, pp. 134–49.

1) 'On lace-making machinery'
2) 'On wire-card making by machinery'
3) 'On cutting furs for pelts'

Part 2: read 12 December 1865 and published *ibid.*, 161–74.

'On the employment of steel for multiplying engravings.' [With an appendix on the Perkins steam gun]

Part 3: read 6 February 1866 and published *ibid.*, 253–9.

'On nail-making by machinery'

Part 4: read 20 March 1866 and published *ibid.*, 297–313.

'Notes on cotton-spinning machinery. Roving frames'

80. Massachusetts Historical Society, Boston. Typescript 'Memoir of Joseph Chessborough Dyer' [1868], 174–5.
81. Most of the genealogical material used in this paper has been taken from family trees in the possession of the Farey family of California.
82. *Gentleman's Magazine*, London, 1804, p. 182.
83. For the Cundys see *Dictionary of National Biography*; Rupert Gunnis, *Dictionary of British Sculptors* (London, 1971), 118; for Samuel Hubert, see H.M. Colvin, *Dictionary of British Architects* (London, 1978), 437.
84. For Cyril Arthur Farey see *The Times*, 8 December 1954; *Journal of the Royal Institute of British Architects*, 1955; Gallery Lingard, *Cyril A. Farey (1888–1954)* (London, 1990) (catalogue of an exhibition of his work). A second exhibition of Farey's work was held at the Gallery Lingard in December 1996.

85. W.S. Mitchell, 'Biographical notes of John Farey, Geologist', in *Geological Magazine*, 1873, 10: 25.
86. Farey, *Treatise on the Steam Engine*, Vol. 1, *op. cit.* (9), 280, 437.
87. G.C. Boase, article 'Bryan Donkin' in *Dictionary of National Biography*.
88. Michael Kassler, article 'John Farey' in *The New Grove* (London, 1980). This is the first edition of *Grove* in which Farey appears.
89. M. Betham-Edwards (ed.), *The Autobiography of Arthur Young* (London, 1898), 372.
90. Information from Dr Hugh Torrens.
91. *Transactions of the Society of Arts*, 1805, XXIII: 12ff.
92. For details of the importance of itinerant lecturers in the diffusion of information about technology during the eighteenth century, see ch. III of Musson and Robinson, *op. cit.* (1).
93. British Library, Additional MSS, 35128, fol. 498.
94. The first volume of Farey's *Derbyshire* has recently been reprinted with a new introduction by Dr Hugh Torrens and with an updated bibliography of the writings of John Farey Sr.
95. Royal Society Archives, John Farey, 'An account of the Great Derbyshire Denudation', MS fol. 1, read to the Royal Society on 21 March 1810.
96. Joan M. Eyles, article 'John Farey', in *Dictionary of Scientific Biography* (New York, 1970–80), Vol. IV, 450–1.
97. John Farey Sr, 'On a Curious Property of Vulgar Fractions', in *Philosophical Magazine*, 1816, 47: 385–6.
98. Sutro Library, San Francisco, Banks MSS, WM1:3, fol. 1.
99. Leonard Bull, *History of the Smithfield Club from 1798 to 1925* (London, 1926), p. 7, 9, 37, 40, 43, 45.
100. Dyer wrote a continuation to his 'Memoir' (*op. cit.* (80)), which had taken his life up to 1818, called 'Patches in the History of Our Own Time', which unfortunately has been lost. Presumably this would have material about the Gamaches' concern.
101. MS family tree in the possession of the Farey family. Typescript history of the American Farey family by the late Arthur Farey, *c.* 1980.
102. PRO: will of Sophia Farey, proved PCC 1843.
103. MS family tree and letters from Kate Hubert in the possession of the Farey family, California.
104. Letter to author from Ian Frazer, dated 27 November 1980.
105. MS letters relating to family history in the possession of the Farey family, California.
106. Archives of University College London: register of students, 1842. He does not appear in the college register, and it would seem he did not complete the course.
107. *Proceedings of the Institution of Civil Engineers*, 1889, 94: 298–300.
108. Letter from Kate Hubert to Mary Farey dated 17 May 1914, in the possession of the Farey family, California.
109. Letter from Kate Hubert to Henry Nathaniel Farey dated 1 May 1898, in the possession of the Farey family, California.
110. Letter from Kate Hubert to Mary Farey dated 13 March 1934, in the possession of the Farey family, California.
111. Letter from Kate Hubert to Mary Farey dated 13 December 1941, in the possession of the Farey family, California.
112. 'William James Russell, 1830–1909', *Proceedings of the Royal Society*, 1911, LXXXIV: xxx–xxxi.
113. Letter to author from Mr Michael Pearson, grandson of Edward's lawyer, dated 28 February 1980.
114. Principal Probate Office, London: will of Edward Farey, proved 4 November 1913.
115. H.W. Dickinson and A.A. Gome, *A Catalogue of Civil and Mechanical Engineering Designs of John Smeaton* (London, 1950), xi–xii. Newcomen Society Extra Publications 5.
116. Kate Hubert's letters in the possession of the Farey family specifically mention all the items listed here except the proofs of the unpublished engravings from the *Treatise on the Steam Engine* and the two drawing books.
117. Letter to the author from C.N. Acheson-Gray of Jansons, Portman Square, London, dated 11 January 1980.

À la Recherche des Ingénieurs Disparus – les Hydrauliciens Néerlandais au Dix-huitième Siècle

MARTIJN BAKKER

SUMMARY

In Search of Vanished Engineers – Dutch Hydraulic Engineers of the Eighteenth Century

Although Russian interest in the recruitment of Dutch hydraulic engineers in the early eighteenth century is well attested, notably at the time of Peter the Great's residence in the shipyards of Amsterdam and Zaandam, later in the century Dutch engineers continued to make their careers in the Russian Empire.

It is unfortunately difficult to trace their careers once they left Holland, and most disappeared leaving no trace. This paper attempts to retrieve some details of this emigration and at the same time throw some light on the history of engineering in Russia at that time.

The creation of canals linking the great Russian rivers that flow towards the Baltic, Caspian, Black and White Seas was early seen as a prime necessity if the economic resources of Russia were to be mobilized. The climate, no less than the distances involved, posed more than ordinary problems for river navigation. Most water-borne traffic passed through the region of Novgorod, and it was here that schemes of river management were first mooted.

In 1783 Jan Pieter Van Suchtelen (1751–1836), a military engineer then aged 32, sought his fortune in the service of Catherine II. As well as taking part in various campaigns, Van Suchtelen advised on canal construction. In 1802 he was appointed director of the Institute of Military Topography, before spending his last years as ambassador to Sweden.

Jacob Eduard de Witte (1723–1809), after distinguished service as Director-General of Public Works in Amsterdam in charge of the planning and execution of fortifications, canals, docks and drainage, was dismissed in 1777 after charges of fraud were brought against him. In 1782 he accepted an invitation to go to Russia and was there responsible for canal construction and work on the port of Riga.

A number of engineers left Holland in 1783, and it is only by chance that some of their names have been recorded. Among them was an artillery officer, Maurits Anne Ver Huell (1761–1799). He decided to join Van Suchtelen, who was, at that time, recruiting engineers to accompany him to Russia. Apart from taking part in various campaigns, he was employed in fortifying Riga and overseeing harbour works at Reval and Liepaja. Another engineer who left Holland in the same year was J.A. Falconi. For reasons yet unclear, although probably having to do with factional strife, another group of engineers emigrated to Russia in 1787. Among them was J.F. De Wolland (d.1818) who already had behind him a distinguished career in the service of the Dutch state. In 1797 he was named Director of Waterways, and worked with great energy in this demanding office. In 1801 his department produced a map of all the navigable waterways of Russia, showing additionally all canals constructed, under construction or projected. One of his last great projects, extending from 1796 to 1810, was to improve the navigation of the Dnieper in the region of the cataracts.

Some other engineers, such as Van der Weide or J.S. Krausz (or Creutz), we know of by name and little else. About Cornelis Redelijkheid, another probable colleague of J.E. de Witte in Russia, somewhat more is known. After working as a military and canal engineer in Holland he left the country for political reasons in 1787, but what he achieved in Russia is unknown.

Why did Catherine II invite all these Dutch engineers to Russia? It seems that Dutch cartographers and hydraulic experts were held in high regard there, and not without reason. It hardly needs to be said, however, that they were not the only foreigners to accept service under the Tsarina. Why they went was to further their prospects or, later on, to escape revolutionary turmoil. And, of course, once the circle had been established in Russia, it was natural for other men of talent to seek to join them.

(Summary by G.H.-S.)

Dès le règne du tsar Pierre le Grand (1682–1725), on constate l'intérêt de la Russie pour ceux qui, en Hollande, s'occupaient des dangers de l'eau et de ses possibilités: constructeurs navals et bâtisseurs de digues, d'écluses et de canaux. En 1696, Pierre le Grand se rendit à Amsterdam et à Zaandam pour se former à la construction navale. À la suite de cette impériale visite, il semble que le vocabulaire naval de la langue russe se soit enrichi de quelque 300 mots d'origine hollandaise. On aurait tort cependant de surestimer le rôle que joua le petit pays baigné par la mer du Nord dans l'évolution des domaines hydraulique et maritime de l'immense Empire russe: des experts de nombreux autres pays y firent aussi des carrières remarquables. Ainsi, pour la construction de canaux, Pierre le Grand embaucha l'ingénieur anglais John Perry, auquel succéda l'Allemand Burchard Christoph von Münnich, en 1723. Au début du dix-neuvième siècle, le Hollandais De Wolland occupait le poste de chef des Voies navigables russes. Il y avait été précédé par Johann Conrad (von) Gerhard, qui était allemand ou autrichien, tandis que son propre successeur serait l'Espagnol De Bétancourt.[1]

Au sein de ce creuset de nationalités que semblent avoir été les Ponts et chaussées, nous examinerons les activités de quelques Hollandais de la fin du dix-huitième siècle et du début du dix-neuvième, puisque c'est à certains d'entre eux que l'on doit les grands travaux réalisés à cette époque.

Le présent article ne peut être qu'une première ébauche. En effet, l'historiographie relative aux ingénieurs hollandais en Russie est très sommaire, à tout le moins aux Pays-Bas. L'historien est confronté à un manque de sources, particulièrement bien reflété par le fait que le sort de nombreux personnages se teintait d'incertitude, même pour leurs contemporains, à partir du moment où ils quittaient la Hollande. Il semble qu'après l'inscription d'annotations telles que 'parti pour Pétersbourg', 'probablement allé en Russie' ou 'en 1787, il est entré au service de la Russie et depuis, on ne sait rien', on refermait le dossier en tout cas en Hollande. Apparemment, ces disparus se refaisaient une vie nouvelle dans une autre patrie. Il est possible que les archives russes contiennent aujourd'hui davantage de renseignements que celles des Pays-Bas. Nous avons bien retrouvé quelques sources contenant des informations sur leurs activités en Russie mais nous nous sommes alors immédiatement trouvés confrontés à un autre problème: celui du manque d'historiographie permettant de replacer ces informations fragmentées dans un cadre plus large. Elles sont également difficiles à vérifier en ce qui concerne, par exemple, l'exactitude de l'orthographe des noms russes, celle de la chronographie, etc.

En dépit de ces lacunes et obstacles, nous nous efforcerons dans cet exposé d'apporter quelques éléments utiles à l'histoire de l'ingénierie russe du dix-huitième siècle et du début du dix-neuvième.

Nous tenterons d'apporter une réponse aux principales questions suivantes: qui étaient ces hommes, que faisaient-ils avant de partir pour la Russie, de quoi étaient-ils capables et pourquoi a-t-on eu besoin d'eux dans l'empire de Catherine II (1762–1796) et de ses successeurs? En associant ces fragments d'informations, issus de sources imprimées et manuscrites ou de documents secondaires, nous essaierons de soulever un coin du voile.

Pour commencer, il faut tout d'abord planter le décor et esquisser brièvement le domaine dans lequel ces Hollandais se sont particulièrement distingués: celui de l'état des eaux russes et de tout ce qui s'y rapporte.

LES EAUX ET VOIES NAVIGABLES EN RUSSIE AU DIX-HUITIÈME SIÈCLE

Sur le territoire russe, on distingue quatre grands bassins hydrographiques. Au nord-ouest, un bassin se déverse dans la mer Baltique. Au nord-est, tous les fleuves se jettent dans la mer Blanche. Dans le sud, ils se jettent dans la mer Caspienne et plus à l'ouest, dans la mer Noire. Dans tous ces réseaux hydrographiques se trouvaient d'importantes réserves en bois, produits agricoles et minéraux. Le problème, comme s'en rendit compte

Pierre le Grand, était de transporter les marchandises d'un bassin à l'autre. Comment relier ces réseaux fluviaux entre eux? Pour le grand Tsar, cette question était d'autant plus cruciale que, depuis la création de St-Pétersbourg au bord de la mer Baltique (en 1703), il devenait de plus en plus nécessaire d'établir une liaison entre la région du Nord et celle du Sud – l'Ukraine et la Crimée (Figure 1).

Les difficultés posées par la navigation intérieure en Russie sont de celles que l'on ne rencontre que rarement dans les régions occidentales de l'Europe. En raison du climat continental, le débit des rivières et des fleuves russes est extrêmement variable d'une saison à l'autre. En hiver, les cours d'eau gèlent et, de ce fait, chaque printemps ramène la menace d'une débâcle destructive. En avril et en mai, le dégel transforme les rivières en torrents tandis qu'en août et septembre, le niveau d'eau peut être très bas: un courant réduit se faufile dans des passages étroits semés d'innombrables étocs et rend la navigation problématique. À l'époque de Pierre le Grand, par ailleurs, pendant les quelques mois où les fleuves étaient navigables, il fallait transporter les marchandises sur de très longues distances et on les regroupait en vastes convois ou caravanes. Vers 1780, il arrivait que 800 bateaux, ou davantage, attendent un niveau d'eau suffisant pour pouvoir passer un obstacle. Certains endroits avaient une réputation redoutable, comme le passage de Vyshniy-Volochek, le lac Ladoga et les cataractes du Dniepr.

La concentration des activités de transport de marchandises sur quelques mois et l'état de plusieurs fleuves entraînaient une autre complication: le trafic était à sens unique et ne pouvait se diriger qu'en aval, vers la mer. Les bateaux ne revenaient jamais à leur point de départ. Il fallait donc construire des milliers d'embarcations pour un seul voyage de plusieurs mois et des régions entières étaient menacées de déboisement.

La région par laquelle transitait la majorité des transports de marchandises russes était la province de Novgorod. En 1764 et 1768, son gouverneur – un Lituanien nommé Jakob Johann Sievers, qui était aussi un familier de la tsarine Catherine II – présenta ses premiers plans pour l'aménagement de plusieurs cours d'eau de la province. Sievers fut nommé directeur général des Eaux pour la région en novembre 1773,[2] et il prit comme adjoint Johann Conrad Gerhard. Ce dernier avait acquis sa notoriété d'ingénieur hydraulicien à la cour de Marie-Thérèse d'Autriche. Au bout de dix ans de labeur intense, Sievers, fatigué, passa la main à son successeur, le comte Bruce. Le peu d'enthousiasme dont celui-ci fit preuve pour sa tâche égalait son manque de compétences. Est-ce pour cela qu'à partir de 1783, au moins deux ingénieurs hollandais furent chargés de plusieurs projets hydrauliques? L'absence de sources relatives à l'organisation du service des eaux et des travaux publics, sous Sievers comme sous le comte Bruce, ne permet pas de donner une réponse précise à cet égard.

À l'époque, le général hollandais Carel Diederik Du Moulin (1728–1793), descendant d'une famille huguenote,[3] était réputé pour son expertise des sièges, de la guerre de mines, etc. En 1777, Charles de Brunswick l'invita, en vain, à rejoindre l'armée prussienne en vue d'y

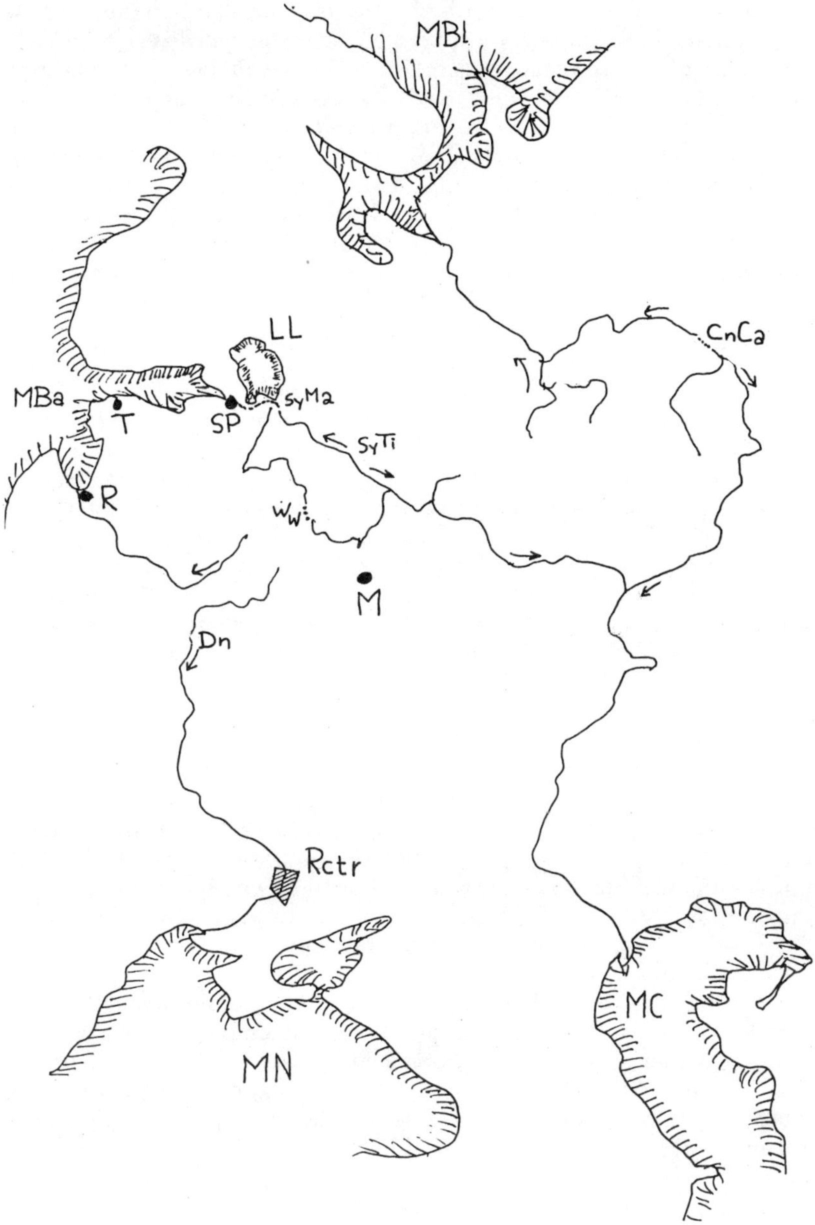

Figure 1 Legende: CnCa = Canal Catherine; Dn = Dniepr; LL = Lac Ladoga; M = Moscou; MBa = Mer Baltique; MBl = Mer Blanche; MC = Mer Caspienne; MN = Mer Noire; R = Riga; Rctr = Région des cataractes; SP = St.-Pétersbourg; SyMa = Système Marie; SyTi = Système Tichwin; T = Tallinn; WW = Vishniy-Volochek.

réorganiser le corps du Génie.[4] Six ans plus tard, c'était au tour de Catherine II de Russie de faire appel à lui. Il semble qu'elle se soit adressée aux États généraux hollandais, disant qu'elle avait besoin de plusieurs 'jeunes officiers du génie qui, avec augmentation de rang', pourraient entrer au service de la Russie. Mais, à nouveau, Du Moulin demeurait à son poste.[5] En revanche, il encouragea l'un de ses aides, Jan Pieter Van Suchtelen, à saisir cette opportunité. Van Suchtelen ne fut pas le seul à se rendre à l'invitation de la Tsarine. Nous allons le voir, ceux qui y répondirent constituait un petit groupe d'ingénieurs ou d'anciens membres de ce corps modeste qui, à l'époque, ne comptait qu'environ 70 officiers de tous rangs.

JAN PIETER VAN SUCHTELEN

Jan Pieter Van Suchtelen appartenait à une famille d'ingénieurs militaires et de cartographes. Son père Cornelis, lui-même fils de militaire, était ingénieur puis fut nommé grand-major de la garnison de Sluis, en Zélande. Son oncle Rochus, nommé lieutenant-colonel du Génie en 1775, fut chef de la Région militaire de la Flandre hollandaise. Il existe encore plusieurs plans de fortifications dus à Cornelis et à Rochus van Suchtelen.[6]

Jan Pieter, né le 2 août 1751 dans la ville fortifiée de Grave sur la Meuse, fut envoyé à Groningue à l'âge de huit ans pour y faire ses études. En 1765, il rédigea en latin un exposé intitulé 'Oratio de Christi doctrinae praestantia', sur la supériorité de la doctrine du Christ. N'ayant pas le goût des études philosophiques, il rentra chez ses parents où son père lui enseigna les mathématiques requises pour lui permettre d'embrasser une carrière militaire. En janvier 1768 (il avait 17 ans), il obtint dans le corps du Génie le grade d'ingénieur-extraordinaire, qui correspond à celui de sous-lieutenant, puis celui de lieutenant-capitaine en 1772. En 1778, il était nommé aide de camp du général Du Moulin, qui commandait le corps du Génie.[7] Lorsque Du Moulin lui transmit l'invitation russe, cet officier de 32 ans y vit des perspectives de carrière beaucoup plus intéressantes que celles que pouvait lui offrir le corps du Génie hollandais.

Le départ de Van Suchtelen, en octobre 1783, fut assez précipité du fait que le 'courrier' qu'il avait trouvé pour l'accompagner pendant son voyage était sur le point de partir. Sa femme et ses enfants le rejoindraient par bateau quelques semaines plus tard. Ils apporteraient avec eux quelques grandes caisses, contenant 'tous ses livres, instruments et papiers'. Parmi ces papiers se trouvaient aussi beaucoup de plans, de dessins et de cartes relatifs à la défense de la Hollande, qui faisaient partie des possessions personnelles de Van Suchtelen.[8]

C'est en 1786 que la légation hollandaise à St-Pétersbourg fait mention de ses activités. Environ un an auparavant, Catherine II avait chargé le lieutenant-colonel ingénieur Van Suchtelen d'étudier les possibilités d'unification de plusieurs fleuves du Nord du pays: la Vytschgeda, la Keltma septentrionale, la Keltma méridionale et la Kama. L'idée était de créer une voie navigable reliant la mer Blanche à la mer Caspienne. Van Suchtelen avait indiqué que l'on aurait besoin d'un canal de 20 werst, avec deux très grandes écluses à sas et deux écluses d'évacuation, dont le coût reviendrait à plus de 400 000 roubles. Il estimait que les travaux demanderaient trois ou quatre ans. À la fin mars 1786, Van Suchtelen partit pour le Nord et Catherine II augmenta à 3000 roubles son salaire de 1500 roubles. Elle lui offrait une prime de 1500 roubles, additionnée de 600 roubles pour ses frais de voyage et de séjour et, en plus, elle décorait l'ingénieur de l'Ordre de St-Vladimir. La guerre avec les Turcs (1787–1792) interrompit les travaux de ce qui serait baptisé le canal Catherine. Les travaux ne reprendraient qu'en 1803, sous la direction d'un autre ingénieur.[9] Entre-temps, on avait rendu hommage au talent de cartographe de la famille Van Suchtelen en nommant Jan Pieter directeur de l'Institut de cartographie, en 1802. Il y succédait au général Arakchev et allait demeurer à ce poste jusqu'en 1808. Trois ans plus tard, Van Suchtelen devenait le premier directeur de l'Institut militaire de cartographie, d'où sortirait en 1814 un relevé cartographique de la Russie en 113 planches à l'échelle du 1/840 000.[10]

Pendant la guerre contre la Suède (1788–1790) et la campagne militaire en Pologne, Van Suchtelen participa à plusieurs sièges et fut chargé des réparations des fortifications aux alentours de Varsovie. Il y gagna un grand nombre de décorations. En 1807, il entama une carrière diplomatique, mais du fait d'une autre guerre russo-suédoise (1808–1809), on fit à nouveau appel à ses compétences d'assiégeant. La grande place forte de Sweaborg fut prise selon son plan, ce qui mit un terme aux hostilités et rendit la Finlande à la Russie. Van Suchtelen fut nommé envoyé spécial en Suède et, en 1812, il signa au nom de la Russie le traité de paix entre la Russie et l'Angleterre. De 1814 à sa mort, en 1836, il vécut en Suède en qualité d'ambassadeur spécial de la Russie et entretint des relations très cordiales avec Bernadotte, qui avait été élu roi de Suède.[11] Deux de ses fils, Paul (1788–1833) et Constantijn (après 1789–?), firent carrière dans l'armée russe.[12]

JACOB EDUARD DE WITTE

Petit-fils d'un colonel de la Garde hollandaise et fils d'un officier d'infanterie qui finit sa carrière colonel-commandant d'un régiment, Jacob Eduard de Witte (vers 1723–1809) était prédestiné à une carrière dans l'infanterie. En 1764, pourtant, il quittait cette arme pour le corps des ingénieurs de fortifications. Il y était nommé lieutenant,[13] au service des États généraux hollandais.

En mai 1772, le poste de directeur général des Travaux publics de la ville d'Amsterdam devint vacant. C'était une position plus prestigieuse que celle d'officier de fortifications. De Witte posa sa candidature et obtint le poste.[14] Depuis 1746, les magistrats y avaient toujours nommé des officiers ingénieurs.[15] De 1500 florins, en 1746, le salaire était passé en 1772 à 4500 florins par an, plus 600 florins pour un cheval et une voiture, montant qui reflète assez l'ampleur de la tâche du directeur général. De Witte était seul responsable des routes hors des fortifications et des voies urbaines, des quais, des écluses, des couloirs et chenaux, des digues des fleuves et des lacs (à l'époque, Amsterdam était bordé sur deux côtés par de grands lacs), des pompes d'épuisement, des ponts mobiles et fixes, et des édifices publics. Sur tout ce territoire, le directeur général avait la responsabilité de l'inspection, de l'entretien et des réparations, et également des nouveaux projets. Il contrôlait l'élaboration de tous les projets et leur exécution, c'est-à-dire l'arpentage et le métrage, la cartographie, le nivellement, l'égalisation, le creusement des fondations, le dessin de plans, les devis et la mise en adjudication. Les magistrats exigeaient qu'il fournisse chaque mois des rapports détaillés sur son travail.

De Witte, au cours de sa carrière d'officier-ingénieur, avait acquis toutes les compétences requises. En novembre 1775, il eut l'occasion de faire la preuve de son sang-froid et de son expertise. Alors qu'il venait d'achever les plans et dessins du nouveau théâtre municipal,[16] une tempête fit monter fortement le niveau des lacs au nord et à l'ouest de la ville. C'est grâce à l'intervention de de Witte, qui supervisa et dirigea jour et nuit la construction de digues provisoires, que la ville dut de n'être pas submergée. Son nom fut salué dans le livre qu'on écrivit un an plus tard sur ces inondations,[17] et il reçut une récompense de 1000 florins. Ce moment de gloire précédait un désastre. Au cours de l'hiver 1776, le comptable du service des Travaux publics faisait savoir aux magistrats qu'il soupçonnait de Witte de fraude. On interrogea alors longuement des dizaines de personnes, puis de Witte eut la possibilité de se défendre et on réentendit enfin les témoins à charge et à décharge. Jacob Eduard de Witte fut démis de ses fonctions en mai 1777.[18] En juin, au désespoir, il suppliait les magistrats de le réhabiliter, lui qui était père de six enfants et dont l'épouse était enceinte du septième. Il se disait victime d'un complot et faisait valoir qu'il avait travaillé jour et nuit pendant cinq ans, six voire même sept jours par semaine, avec un seul assistant et quelques subalternes. N'y avait-il vraiment pas dans la ville un poste pour lui au nouveau Directorat général qui avait été créé immédiatement après sa démission?[19] Les magistrats se montrèrent implacables et de Witte quitta la ville. À partir de cette date, nous devons nous reporter aux mémoires manuscrites de son fils, Jacob Eduard de Witte le jeune, qui rédigea son autobiographie en 1826.[20] Il semble que son père ait subsisté quelques années durant en donnant des avis sur les travaux hydrauliques entrepris dans les provinces de Gueldre et d'Utrecht. La mort de son épouse, vers 1781, lui porta un nouveau coup, se souvient son fils, jusqu'à ce que de Witte soit invité en 1782 par le prince Galitzine, ambassadeur russe à La Haye, et 'son ami le général Du Moulin' à s'engager dans l'armée russe au grade de colonel du Génie, affecté à la garnison de Riga. Nous verrons plus loin que les mémoires du fils de Witte, qui était demeuré en Hollande, manquent parfois d'exactitude en ce qui concerne les activités de son père. Elles fournissent toutefois des détails utiles. 'Un jour, écrit le fils, mon père fut chargé d'effectuer l'union de la Dvina avec la Volga et il consulta ses amis et compatriotes, le colonel Van Suchtelen, le major Redelijkheid, le capitaine Krausz et le capitaine Van der Wijde.' Pour l'exécution des travaux, il avait été décidé de faire venir de l'étranger les instruments et outils nécessaires, ainsi que des ouvriers habiles à bâtir des écluses. De Witte écrivit à ses anciens amis à Amsterdam et une cinquantaine d'ouvriers hollandais se rendirent ainsi en Russie. Par ailleurs, les outils étaient arrivés en temps voulu. La tsarine Catherine II montra sa satisfaction du déroulement des opérations en élevant de Witte au grade de brigadier-général et le décorant de l'Ordre de St-Vladimir. Les travaux furent achevés en trois ans et de Witte fut nommé lieutenant-général et comte de l'Empire.[21] Comparons cette version des mémoires du jeune de Witte avec une autre source, à savoir les rapports de la légation hollandaise à St-Pétersbourg. En mars 1786, l'ambassadeur de Hollande écrivait que le major de Witte était chargé d'étudier la possibilité de relier deux fleuves, la Vitegra et la Kauvscha.[22] Les plans et les cartes étaient prêts à la fin du mois de mars mais de Witte fut envoyé à Riga pour y superviser un projet dont il s'était déjà occupé auparavant. On avait besoin à Riga d'une très grande écluse à sas, par laquelle les bateaux de gros tonnage pourraient passer de la Dvina occidentale à la fosse de la ville. De là, ils seraient dirigés vers un vaste bassin, qui n'était pas encore construit, où ils pourraient hiverner, loin de la débâcle dangereuse de la Dvina.[23] En octobre, l'ingénieur-major de Witte recevait l'Ordre de St-Vladimir. Après cet épisode, de Witte n'est plus mentionné dans les rapports de l'ambassadeur. Dans les mémoires de Sievers, on lit qu'en 1797, de Witte avait été chargé par le tsar Paul I de la jonction de la Dvina avec l'Aa, près de Riga, et qu'il avait dessiné le tracé d'un canal reliant Kovno à Mitau.[24] Il mourut en mai 1809. Ses enfants, qui l'avaient accompagné en Russie, firent de belles carrières dans leur nouvelle patrie: sa fille épousa le comte Chirnitchiev, conseiller d'État; quant à ses fils, Frederik Hendrik devint capitaine du Génie et aide du général Van Suchtelen; Pieter Lodewijk fut avocat auprès de la Haute cour de justice russe; et Adriaan Kasper était colonel d'artillerie lorsqu'il fut tué au siège d'Oczakov contre les Turcs.[25]

QUELQUES AUTRES: VER HUELL ET FALCONI

On ignore le nombre exact d'ingénieurs ayant quitté la Hollande en 1783. Ce n'est que par hasard que l'on en retrouve les traces. Par exemple, Sievers se contente de mentionner dans ses mémoires l'arrivée, sous le règne de Catherine II, de quelques Hollandais dont les noms deviendraient célèbres plus tard: de Witte, Van Suchtelen, Verhuël.[26] Ce dernier est probablement Maurits Anne Ver Huell (1761–1799)[27] qui, comme Van Suchtelen et de Witte, appartenait à une famille militaire.[28] Maurits montra des dispositions pour le dessin et l'arithmétique, et lorsqu'il fut âgé de 14 ou 15 ans, son père, qui était maire de Doetinchem, le mit dans une école d'artillerie à Bois-le-Duc. Au bout de quelques mois, Maurits Ver Huell obtenait en octobre 1777 son brevet d'ingénieur-extraordinaire au service de l'État. On l'affecta à la garnison de Zutphen où il fut chargé des petites réparations aux bastions mais, bientôt, l'ennui le gagna. Il saisit sa chance alors que Jan Pieter Van Suchtelen cherchait des ingénieurs pour l'accompagner en Russie. Avec la permission de ses parents, le lieutenant Ver Huell partit avec lui et fut immédiatement promu capitaine. Au cours de la guerre russo-suédoise de 1788–1790, son courage et sa présence d'esprit lui valurent plusieurs décorations. Le tsar Nicolas I, visitant les chantiers navals de Rotterdam en 1798, rencontra un neveu de Ver Huell et lui parla des fortifications de Riga que son oncle avait bâties et dont un bastion portait même le nom. En dehors des fortifications, Ver Huell est également connu pour des constructions à Reval (Tallinn), à Liepaïa, au sud de Riga, et pour la construction de Port-Baltique, à l'ouest de Tallinn. Son supérieur était Jacob Eduard de Witte avec lequel il entretenait aussi une solide amitié. En 1799, de Witte écrivit à la famille de son ami que le lieutenant-général Maurits Ver Huell, après plusieurs mois d'une 'hypocondrie' proche de la démence et de fortes fièvres, était décédé entre le 15 et le 26 novembre 1799.

D'un seul autre ingénieur, nous savons qu'il est parti en 1783. Il s'agit de J. A. Falconi, ingénieur-extraordinaire au corps du Génie hollandais à partir de 1777. Il était affecté à la garnison de Berg-op-Zoom.[29] Sur la liste des officiers de 1783, son nom a été rayé, ce qui indique qu'il avait quitté ce corps. Ce n'est qu'en 1798 que nous retrouvons sa trace. Il menait une existence difficile dans une petite propriété de Lituanie. Depuis qu'il avait quitté la Hollande, Falconi n'avait été employé qu'à des actions militaires sous les ordres du général Souvorov et n'avait jamais pu utiliser ses compétences d'ingénieur hydraulicien. Lorsque Souvorov tomba en disgrâce, il fut impossible à Falconi, comme à beaucoup d'autres, d'obtenir un autre poste. Alors qu'il se trouvait dans le dénuement, il fut retrouvé par un ancien collègue et compatriote qui se souvint de ses qualités et lui offrit un emploi dans la construction des canaux.[30] Le sauveur de Falconi s'appelait François De Wolland.

LE GROUPE DE 1787: DE WOLLAND, CREUTZ ET REDELIJKHEID

Pour des motifs demeurés obscurs, un deuxième groupe d'ingénieurs se rendit en Russie en 1787. Il est possible que les problèmes politiques en Hollande à cette période aient joué un rôle: la République des Provinces-Unies était pratiquement en révolution et la situation devenait plutôt difficile pour les partisans de la maison d'Orange et du stathouder Guillaume V. Les 'vrais Républicains' formaient de petits groupes armés et menaçaient les Orangistes. Parmi beaucoup d'autres, le capitaine-ingénieur De Wolland fut traité de manière insultante. Les États généraux l'avaient chargé de dessiner une carte de la frontière, dans la province d'Overijssel, à l'est du pays. Il fut arrêté sans préavis par un groupe de révolutionnaires en armes parce qu'il était soupçonné d'espionnage au profit des Orangistes. Tous ses biens lui furent confisqués.[31] Nous ne possédons pas d'informations concernant ses préférences politiques mais nous savons que, le 9 novembre 1787, De Wolland rencontrait Kalitchev, l'ambassadeur russe à La Haye, pour signer un contrat par lequel il entrait au service de la Russie en qualité de major dans le corps des Ingénieurs, avec un salaire de 1200 roubles par an.[32] Il était alors immédiatement envoyé vers le sud de la Russie pour prendre part à la guerre contre les Turcs (Figure 2).

Ce n'était pas le premier grand voyage qu'entreprenait De Wolland. On ne sait pas grand chose de ses origines, mais il semble qu'il venait de la région bruxelloise. En 1779, il se trouvait en mission cartographique au Suriname, en Amérique du Sud, où, à partir de 1781, il élaborait le plan de défense de cette colonie hollandaise contre les Anglais. La Hollande et l'Angleterre étaient en guerre depuis 1780 et les Hollandais craignaient que leur ennemi ne s'attaque à leurs territoires d'outre-mer. Afin d'organiser une défense efficace contre une telle opération, De Wolland et quelques autres entreprirent de dresser de nouvelles cartes topographiques de la côte du Suriname et de ses places fortes. Le premier plan ayant été rejeté par les États généraux en Hollande, De Wolland envoya en juillet 1784 une nouvelle proposition contenant, entre autres, l'extraordinaire idée d'une barrière flottante de 2410 mètres de long, sur un mètre de diamètre, destinée à empêcher les navires ennemis d'accéder au fleuve Suriname. De Wolland se trouvait à La Haye fin 1784 et, en janvier 1785, le général Du Moulin, qui le tenait en estime, le chargeait de dessiner la carte militaire de la région Est du pays. C'était là qu'en 1787, les révolutionnaires lui menaient la vie dure.[33]

En Russie, comme nous l'avons indiqué, De Wolland se trouvait sur le front Sud. Il s'occupait des fortifications et il fut promu au grade de brigadier en 1795. Vers la fin de l'année suivante, il se rendait à St-Pétersbourg pour y prendre la tête du Département des Fortifications. Broyé par la bureaucratie et les décisions arbitraires des autorités, De Wolland était malade et perdait tout espoir de rétablir sa situation. Il souhaitait s'installer à Vienne où des personnages importants lui avaient proposé de poursuivre sa carrière. C'est à ce moment-là que Jakob Johann Sievers fut mis au courant de l'existence de ce brillant ingénieur.

Malgré son âge – il avait déjà plus de 70 ans –, Sievers s'était fait nommer directeur en chef des Eaux en février 1797. Avec énergie, il reprit les activités qui avaient été arrêtées sous la direction de son prédecesseur immédiat. En tout cas, la collaboration entre De Wolland et Sievers donnait de bons résultats. Après la mort de Sievers, en 1800, De Wolland fut en mesure de poursuivre la gestion de plusieurs grands travaux. Dans l'entourage du Tsar, il s'était acquis la bonne réputation de travailler très vite et d'avoir l'oeil aussi bien sur les détails que sur l'ensemble des plans. Malgré le rhumatisme très douloureux dont il souffrait, De Wolland faisait des inspections dans des régions éloignées et peu hospitalières. En 1801 était réalisée sous sa direction une carte au 1/420 000 des voies navigables de la Russie, qui indiquait sur 18 cartouches les plans de tous les canaux alors terminés, en projet ou en construction.[34] Parmi les autres travaux

Figure 2 François de Wolland, ingénieur hollandais, auteur du premier projet d'amélioration des cataractes du Dniépr.

qu'il réalisa, citons: un canal sans écluses près du lac Onega; des contributions importantes au 'système Marie', une série de fleuves canalisés (182 kilomètres) et de canaux (311 kilomètres, 38 écluses) entre la Volga et St-Pétersbourg; le plan du 'système Tichwin', un canal de sept kilomètres qui mène la navigation intérieure sur un partage des eaux de 160 mètres, avec six écluses et 50 barrages. Entre 1796 et 1810, De Wolland s'occupa en outre d'un projet d'amélioration des cataractes du Dniepr.[35] Son talent et son dévouement énergique lui valurent enfin la direction des voies navigables. Il y succédait au prince Georges Paul Oldenbourg, cousin et beau-frère du tsar Alexandre I, qui avait montré une certaine compétence pour ce type de travail.[36] L'ingénieur allemand Wiebeking se souvenait qu'en 1817, De Wolland, 'un homme sincère et dévoué aux intérêts publics, présenta au Conseil des ministres un tableau de l'état effrayant des voies navigables intérieures et des suites funestes qui devaient en résulter (...), qui menaçaient la navigation vers St-Pétersbourg d'une interruption totale'.[37] De Wolland mourut en 1818.

Nous avons pu retrouver la trace de quelques uns des 'amis et compatriotes' de son père que nommait de Witte le jeune dans ses mémoires. Nous ne savons rien de Van der Wijde, en dehors de la lettre qu'adressa Van Suchtelen à Maurits Ver Huell en 1794, où il lui demandait 'de faire ses compliments à M. van der Weide'.[38] Un autre nom était celui de Krausz, probablement l'orthographe incorrect de Creutz. Johan Samuel Creutz fut l'assistant de Jacob Eduard de Witte à Amsterdam et, après 1777, l'un de ses successeurs. Né vers 1747 à Berlin, Creutz s'inscrivit en septembre 1766 comme étudiant en philosophie à l'université de Leyde. Cinq plus tard, il devenait l'assistant de M. Rauws, le directeur général des Travaux publics à Amsterdam. En 1772, de Witte succédait à Rauws et Creutz demeurait assistant, principalement chargé de l'architecture. En dehors de ses activités pour la ville d'Amsterdam, Creutz exerçait sa profession d'architecte pour son propre compte et il reste encore quelques-uns de ses bâtiments à Leyde et ailleurs. Depuis 1783, par ailleurs, il faisait partie de la Commission pour la mécanique et la chimie de la Société hollandaise des sciences.[39] Après la démission de de Witte, Creutz présenta un mémoire sur la réorganisation des travaux publics et, comme ses idées avaient été acceptées, il fit partie du nouveau directorat composé de trois personnes. Il devenait responsable de la division 'Hydraulique', relative à tout ce qui concernait la circulation des eaux, le dragage, la construction de quais, etc. Ce poste ne correspondait probablement pas aux ambitions de Creutz, celui de directeur de l'architecture civile étant plus prestigieux et plus conforme à ses intérêts personnels. À l'issue d'autres réorganisations du service et de quelques conflits avec la Magistrature et ses collègues (Creutz ne paraît pas avoir été doté d'un caractère très docile), il donnait son congé le 20 juin 1787. Dès le 10 juillet, il partait 'pour Pétersbourg', comme il est noté dans les archives de l'Église wallonne. Il semble que Creutz ait pris sa décision avec très peu de concertation. Ses supérieurs furent mis devant le fait accompli de son enrôlement en qualité de major du Génie en Russie, et il laissait en Hollande sa femme et sa fille alors âgée de 13 ans.[40]

Nous avons pu retrouver des détails biographiques de Redelijkheid, un autre collègue de de Witte en Russie. Il s'agit probablement de Cornelis Redelijkheid (ou Reedelijkheid), hydraulicien compétent, mécanicien et expert en fortifications, dont le *Nieuw Nederlandsch Biografisch Woordenboek*[41] disait qu'il avait 'peut-être fui en 1787 (pour des raisons politiques) comme tant d'autres, mais il nous manque toute information sur le reste de sa vie'. En septembre 1788, la légation hollandaise à St-Pétersbourg écrivait aux États généraux que, pendant le siège d'Oczakov, le colonel Redelijkheid, ayant le grade de brigadier dans le corps du Génie, était mort des suites de fièvres chaudes.[42] Fils d'un maçon de Rotterdam, Cornelis Redelijkheid naquit en 1728. Il entra au service des Eaux des États de Hollande en 1761. Il publia un grand nombre d'ouvrages sur les inondations, la maçonnerie et les fortifications, sur les moyens de mieux lutter contre les crues, et sur quelques doctrines du protestantisme. Par ailleurs, il inventa plusieurs machines de dragage et on lui doit le plan d'une écluse à portes à coulisses qui lui valut une récompense de 1000 ducats d'or de la part des États de Hollande. Pourtant, il faudrait attendre 1879 pour que soit construite aux Pays-Bas la première écluse selon ce plan!

Vers 1784, Redelijkheid avait obtenu le grade de colonel-ingénieur et il semble qu'avant de partir pour la Russie, il ait été inspecteur des fortifications et de l'état des eaux à Bréda.[43]

CONCLUSION

Après tous ces petits détails biographiques, nous tenterons, à l'aide de quelques questions, de résumer les premières conclusions de cette exploration.

Pourquoi Catherine II avait-elle invité des Hollandais? Il semble qu'en Russie, on tenait en haute estime les aptitudes des ingénieurs hollandais tant en ce qui concerne la chose militaire que sur le plan de la cartographie et de l'hydraulique. Nous ignorons si, à la même époque, on recherchait également des experts dans d'autres pays et nous ne sommes donc pas en mesure de dire si les Hollandais constituaient un groupe spécial et dominant dans les ponts et chaussées russes. Par exemple, les Hollandais offraient-ils vraiment à l'époque des avantages sur les autres nationalités en ce qui concerne la formation? Il n'existait pas aux Provinces-Unies d'école spéciale pour les officiers-ingénieurs. Ceux-là étaient tous formés par une éducation privée, par leur père ou par d'autres membres de leur famille, souvent militaire, comme on peut le déduire des biographies de Van Suchtelen, de Witte et Ver Huell. À maintes occasions, le général Du Moulin remit aux États généraux des notes pour protester contre le manque de formation de beaucoup d'ingénieurs du corps du Génie. Dès leur entrée dans l'armée, certains jeunes officiers se formaient en autodidactes et mettaient un point d'honneur à se tenir au courant des développements internationaux dans leur domaine en matière de cartographie, d'hydraulique, de mathématiques et, surtout, de géométrie. Mais à cet égard, les Hollandais n'étaient pas nécessairement supérieurs à leurs homologues d'autres nations. Ils avaient peut-être l'avantage d'être nés dans un pays où la construction de digues et d'écluses de toutes sortes, ainsi que la lutte séculaire contre l'eau en général, avaient donné naissance à une somme extraordinaire de connaissances, fondées sur l'expérience et la pratique, que l'on trouvait rarement ailleurs.

Les protestations du général Du Moulin suscitent une autre question. Pourquoi ces hommes, dont nous venons d'esquisser le portrait, avaient-ils un tel désir d'émigrer? Le corps du Génie des Provinces-Unies n'offrait pas vraiment les perspectives d'une brillante carrière. Il ne comptait que de 70 à 80 (sous-)officiers et aucun soldat. Les salaires étaient peu élevés et les possibilités de promotion y étaient restreintes. Pour ceux qui, encore jeunes mais déjà talentueux, y obtenaient un poste, l'avenir n'était guère prometteur. C'est pour cette raison que de Witte quitta le Génie pour devenir directeur général des Travaux publics à Amsterdam. Et Van Suchtelen savait que son poste d'aide du général Du Moulin, qu'il avait obtenu en 1778 à l'âge de 27 ans, pourrait bien être son bâton de maréchal s'il ne trouvait pas un autre emploi ailleurs.

Il reste encore d'autres aspects plus personnels que nous nous contenterons de citer, toute analyse ne pouvant être qu'hypothétique. Citons d'abord le climat politique des années 1780, raison pour laquelle De Wolland et Redelijkheid répondirent si facilement à l'invitation russe. Par ailleurs, il est vraisemblable que les ingénieurs hollandais, en poste en Russie, recevaient de temps à autre des demandes d'emploi de la part de leurs anciens collègues en Hollande. Ainsi, petit à petit, les Hollandais formaient en Russie un petit réseau dont les réalisations donnaient toute satisfaction à la Tsarine et à son entourage, ce qui contribuait à renforcer la réputation d'être de bons hydrauliciens que l'on accordait depuis si longtemps à tous les Hollandais.

Notes et Références

1. Karl Ludwig Blum, *Ein Russischer Staatsmann: Des Grafen Jakob Johann Sievers Denkwürdigkeiten zur Geschichte Russlands*, 4 vols (Leipzig/Heidelberg, 1858), I, 418–421, et II, 4.
2. Blum, *op. cit.*, I, 267, 397, et IV, 419.
3. Voir H.J. Wolf, 'Luitenant Generaal Du Moulin, vestingbouwkundige uit de 18e eeuw', *Genie*, 10 (1960), 131–136.
4. Udo von Bonin, *Geschichte des Ingenieurkorps und der Pioniere in Preussen*, vol. I (Berlin, 1878).
5. En 1788, l'ambassadeur de Hollande à St-Pétersbourg écrivait aux États généraux qu'il se répandait une rumeur tenace disant que le général Du Moulin entrerait au service de la Russie à très bref délai en qualité de chef du corps du Génie, avec un salaire de 12 000 roubles et le don de 500 paysans à Livonie, d'une valeur de 100 000 roubles. Algemeen Rijksarchief (Archives nationales, La Haye), archief Staten Generaal, legatie Rusland, inv. n° 154, lettre du 27 mai 1788.
6. Voir N. Van Suchtelen, 'Kartografen uit de familie Van Suchtelen in de achttiende eeuw', *Geodesia*, 26 (1984), 256–259.
7. *Naem-Registers van Officieren*, 1760–1793.
8. Algemeen Rijksarchief (La Haye), 1e Afdeling, archief Raad van State, inv. n° 1117, okt. 1783.
9. Pour plus de détails sur ce canal et bien d'autres, voir E. F. de Hoerschelmann, 'Aperçu historique du développement des voies navigables de l'empire de Russie', *IVe Congrès international de navigation intérieure* (La Haye, 1894). Hoeschelmann ne fait pas mention de Van Suchtelen.
10. A. Tutein Nolthenius, 'Werken van Nederlanders op Waterbouwkundig gebied in Europeesch Rusland', *Weg-en Waterbouw*, 4 (1944), 58–62.
11. On trouve un très grand nombre de lettres (traduites en suédois) de Van Suchtelen à Bernadotte dans A. Ahnfelt, *Tva krönta rivaler*, 2 vols (Stockholm, 1888).
12. *Algemeene Konst- en Letterbode*, (1836), 242–247, 258–263. Il s'agit d'une nécrologie de Van Suchtelen, empruntée à la revue suédoise *Minerva*, n° 8 et 9, 1836; Van der Aa, *Biographisch Woordenboek*, vol 17–II (Haarlem, 1874), 'Jan Pieter Van Suchtelen', 1066–1069, 'Paul Van Suchtelen', 1069.
13. Le 13 juillet 1764: *Naemregister der Heeren Officieren*, 1771, garnison de Woudrichem.
14. Gemeentearchief Amsterdam, archief Thesaurier Ordinaris (arch. 5039), inv. n° 25, fol. 204.
15. Voir Gemeentearchief Amsterdam, H. W. Werkman, *Inleiding bij Inventaris 5040, Archief Stadsfabriekambt 1532–1811*, 9.

16. En mai 1772, l'ancien théâtre avait été détruit par un incendie, au cours duquel le prédécesseur de de Witte avait trouvé la mort.
17. J.H. Hering, *Bespiegeling over Nederlandsch Waternood tusschen den 14den en 15den November 1775*, 2 vols (Amsterdam, 1776). De Witte souscrivit à deux exemplaires.
18. Gemeentearchief Amsterdam, archief Thesaurier Ordinaris, inv. n° 25, fol. 338.
19. Gemeentearchief Amsterdam, (arch. 5031), archief Burgemeesters, ambten en officieren, voorl. inv. n° 1423.
20. Algemeen Rijksarchief (La Haye), 1e Afdeling, aanwinsten 1879 A XII.
21. Algemeen Rijksarchief (La Haye), 1e Afdeling, aanwinsten 1879 A XII, p. 12–18.
22. Il s'agissait probablement du canal Marie, indiqué sur le plan.
23. Algemeen Rijksarchief (La Haye), 1e Afdeling, Staten Generaal, legatie Rusland, inv. n° 154, lettre 4/15 mars 1786.
24. Blum, *op. cit.* (1), IV, 491, 541.
25. Algemeen Rijksarchief (La Haye), 1e Afdeling, aanwinsten 1879 A XII, p. 20. On ne sait pas avec certitude si le célèbre comte S. J. Witte (1849–1915) était un descendant de la famille de Witte.
26. Blum, *op. cit.* (1), IV, 480.
27. L. Turksma, 'Een Gelders vestingbouwer in Rusland', *Kronyck van Deutekom en Salehem*, 13 (1989), 57–59.
28. Son frère Carel Hendrik Ver Huell (1764–1845) aurait été l'un des amiraux de Napoléon.
29. *Naem-Registers van Officieren*, 1777–1783.
30. Blum, *op. cit.*, IV, 487.
31. *Nieuwe Nederlandsche Jaerboeken* (1787), 2177.
32. Blum, *op. cit.*, IV, 480.
33. Pour cette période de la vie de De Wolland, et ses plans dans les archives néerlandaises, voir *Caert-Thresoor. Tijdschrift voor de Historische Kartografie*, 1 (1982), 6 et 4 (1986), 46–50; A. Tutein Nolthenius, *op. cit.* (10), *Weg- en Waterbouw*, 6 (1946), 97–104; Blum, *op. cit.* (1), IV, 480, dit qu'avant d'aller en Russie, De Wolland avait visité l'Angleterre 'dont il connaissait très bien les canaux'. Nous n'avons trouvé aucun autre renseignement sur ce point.
34. La Bibliothèque de St-Pétersbourg en possède un exemplaire.
35. Pour de nombreux détails sur le projet de De Wolland, voir V.E. Timonov, 'Les Cataractes du Dniepr', *IVe Congrès international de navigation intérieure* (La Haye, 1894).
36. William L. Blackwell, *The Beginnings of Russian Industrialization, 1800–1860* (Princeton, 1968), 266.
37. Cité par A. Tutein Nolthenius, *op. cit.*, *Weg- en Waterbouw*, 6 (1946), 98 (probablement extrait de C. F. de Wiebeking, *Analyse descriptive, historique et raisonnée des monuments de l'Antiquité, des édifices les plus remarquables du Moyen Âge et des derniers temps, et des constructions hydrauliques en Europe*, vol. V (Munich, 1840).
38. Rijksarchief Gelderland (Arnhem), archief Familie Ver Huell (archief 0490), inv. n° 35.
39. C.A. van Swigchem, *Abr. van der Hart, 1747–1820, Stadsbouwmeester van Amsterdam* (Amsterdam, 1965), 9.
40. Gemeentearchief Amsterdam, (archief 201) Waalse Gemeente, inv. n° 50E, 10 juli 1787; Gemeentearchief Amsterdam, H. W. Werkman, *op. cit.* (15), 9 et suivantes; R. Meischke, 'De vernieuwing van Rhijnhof en de architect Johan Samuel Creutz', *Leids Jaarboekje*, 51 (1959), 105–124.
41. *Nieuw Nederlandsch Biografisch Woordenboek*, vol. VIII (Leyde, 1930), 1266–1268.
42. Algemeen Rijksarchief (La Haye), archief Staten Generaal, legatie Rusland, inv. n° 154, lettre du 20 septembre 1788.
43. Voir aussi *De Nederlandsche Leeuw*, 91 (1974), 401, où est indiqué que, dans la famille de sa femme, Cornelis Redelijkheid était connu comme étant colonel d'artillerie.

Book Reviews

S.A. Jayawardene, *The Scientific Revolution: An Annotated Bibliography*, Locust Hill Press, West Cornwall, CT, 1996, pp. xvii + 383; 13 ills.

The first essential of a bibliography is accuracy in description. The descriptions here are terse – dimensions of books are not given – but may be relied on; Mr Jayawardene is a meticulous bibliographer.

Then, in some cases, the next requirement is completeness. There is little point in a bibliography of Robert Hooke if that author's printed works are incompletely listed in it. Not all bibliographies aim to be wholly complete, however, and the 1646 items listed here could not conceivably provide a *complete* bibliography of the scientific revolution. The present work is explicitly limited to secondary sources, excluding therefore all editions of original texts whether old or new, edited or translated. One has only to consider such a work as John Fulton's *Bibliography of Robert Boyle* (1947, 1961) to see how vast such a work would be. Nor does Mr Jayawardene pretend to list every journal article that might in some way be relevant to his subject; it is (the author writes) 'a key to the printed literature of the history of science over the 250 years that span the so-called Scientific Revolution'. It is therefore not merely concerned with the *concept* of a scientific revolution – and in fact the only piece from the pen of Thomas S. Kuhn listed here is his book of 1957, *The Copernican Revolution.*

The primary arrangement of the book is by subjects, starting with reference books and ending with the scientific revolution in general; and each section subdivides (at least) into Bibliographies and Monographs plus Articles. If a work has important bibliographical content (like [832] F.R. Johnson, *Astronomical Thought in Renaissance England*, 1937) it may be listed as a bibliography only. Individual items are chronologically arranged. Each item has a quotation or a comment saying what it is about, varying in detail. Sometimes, but not always, the names of contributors to symposium volumes (or some of them) with the titles of their papers are given, sometimes not. Very often reviews of a work are recorded. Systematic searching was limited to works printed before 1990.

> The material in the bibliography consists mainly of bibliographies, encyclopedias, library catalogues, monographs, source books, articles in periodicals, exhibition catalogues, proceedings of symposia, and essay reviews. As the bibliography is restricted to secondary sources,

> collected works of scientists have been included only in exceptional cases. (Introduction, p. xii)

I return, reluctantly, to the question of inclusion/exclusion in this volume, no clear principles being stated. I can understand that the compiler saw fit to exclude highly specialist articles of narrow scope, for example Alexandre Koyré and I. Bernard Cohen on 'The Case of the missing *Tanquam*', though both are distinguished contributors to the historiography of the scientific revolution, whose other works are elsewhere noted. But why omit *every* journal article by these authors? Every original book by Charles Singer is excluded, and one might therefore guess that all general histories were passed over: but J.L.E. Dreyer's *History of the Planetary Systems* (1906) is listed, as is A.N. Whitehead's *Science and the Modern World.* Not only is no place found for such an old classic as Arthur Lovejoy's *Great Chain of Being* but T.S. Kuhn's *Essential Tension* is absent too. I could not find Jacques Roger, *Les Sciences de la vie dans la pensée française du XVIIIe Siècle*, although perhaps this latter work appeared too late to be recorded. All these works, I would argue, are relevant to this bibliography.

The reader should understand, however, that this is a selective (and perhaps, it might even be added, subjective) bibliography of the topic, and that the author intends the reader to make use of the more detailed bibliographies listed in it to supplement the books and papers separately noted. To this extent the reader may regard Mr Jayawardene's book, a labour of love and one of great usefulness, compiled over many years, as a 'bibliography of bibliographies'.

A. Rupert Hall

Gunpowder: The History of an International Technology. Edited by Brenda J. Buchanan, Bath University Press, 1996, xxvii + 403 pp., index, illustrated.

The history of such an important aspect of technology as gunpowder has, perhaps surprisingly, been singularly poorly researched and published over the years. The two major works are the synthesis by Partington published posthumously in 1960 and the volume on gunpowder in the monumental history of Chinese technology by Needham and others published in 1986. Both deal extensively with the seemingly intractable problem of separating the facts from the myths of where and when gunpowder was first discovered and used. As much research and many publications, both before and since these works, have concentrated on the same issues, the resultant corpus is unevenly slanted towards the early period. Research on the later development of gunpowder technology, its influence on other industries, and the economic factors of its manufacture and trade has tended to lag behind, and only in the field of its use and development for warfare have there been any advances. After such a dearth of material comes this volume, the result of a conference organized by the

International Committee for the History of Technology, ICOHTEC, in 1994 at the University of Bath, during which 'The manufacture and marketing of gunpowder' formed one of the four main themes. The sessions of ICOHTEC were organized by Brenda Buchanan, who has edited the papers for publication.

The papers are arranged in three sections, by theme rather than in a strict chronology. Intriguingly, a paper by Robert Howard, on the evolution of powder from an American perspective, comes first, providing an overview of the processes involved in the manufacture of powder as well as information concerning its production in America in the late modern period. Papers on Chinese rockets and the role of the Mongols in the spread of powder technology in Asia lead to what are perhaps the two most important papers in this section. The first, by Gerhard Kramer, deals with the so-called firework books of the fifteenth century. Kramer brings the discipline of the scientist to the recipes and formulas, and provides a rare new light into the chemistry of early gunpowder mixtures. His paper is marred only by an uncritical acceptance of the role of the fabled Niger Berchtoldus in the discovery of powder. The second paper, by Bert Hall, looks as the problems of why early gunpowder mixtures were so prone to 'spoiling' and how the changes in powder technology, especially in the area of grain size, affected the design and effectiveness of early artillery. In the former Hall builds on the work of Kramer, who suggests that early saltpetre was not potassium nitrate but the very highly hygroscopic calcium nitrate.

This section is completed by papers on a little-known firework book in the collection of the Royal Armouries, a short review of the early use of gunpowder and gunpowder weapons in Europe, and a fascinating account of the manufacture of fireworks in Galicia in the twentieth century.

The second section deals with the development and diffusion of gunpowder technology and has two main themes: the geographic spread of powder-making; and the use of powder in mining. In the former, papers include surveys of powder-making in Hungary, Sweden, Spain, Norway, England and France. The Swedish experience of centralized control and organization, described by Bengt Åhslund, provides a great deal of hard evidence for the methods and production capacity of saltpetre-makers in the late seventeenth century. From Norway comes information about the consumption of powder in the silver-mines of Kongsberg and about prices and production from the eighteenth and nineteenth centuries. The very useful survey of gunpowder mills in Spain from the sixteenth to the eighteenth centuries also covers the transfer of technology to the New World. The problems of quality control and the use and (possible) abuse of power at government level in the awarding of contracts in the Bristol area of England are highlighted by Brenda Buchanan, providing a reminder that politics and prejudice can be influential in decision-making. The French experience, described by Patrice Bret, of the move from private to state-run industry in the years around the turn of the nineteenth century highlights the contribution not only of gifted individuals but of the cultural background of the state. The latter theme of the use of gunpowder in mining is often entwined with these surveys of powder-making and production levels. The paper by Heinz Wild offers a brief overview of the origins and diffusion of powder in mining from the early sixteenth century in the Tyrol to the rest of Europe.

The last section deals with powder-making in its prime, which is taken as meaning its final flowering in the nineteenth century. Included in this section is a paper by Seymour Mauskopf on the early scientific study of gunpowder, which clearly indicates that it has been at the forefront of scientific experiment and theory. Interestingly, papers on the introduction of powder for mining in Australia give a sense that here the technology was being learned anew. In contrast, two papers on the industry in Germany, in the Kingdom of Hanover and at Spandau, highlight the very high level of sophistication and specialization achieved by the European industry in the nineteenth century. The two papers on the industry in Ireland, specifically at Ballincollig, County Cork, and at Waltham Abbey in England indicate that the industry is at last being taken seriously by industrial archaeologists and that preservation of the sites is as important as it is for other industrial complexes. The final paper brings us a brief survey of the introduction of smokeless powders and the age of modern propellants, which have all but made the manufacture of black powder extinct.

Because of the number of papers, 24 in all, many of them are short and present summaries of work published elsewhere, often in another language. This brings me to perhaps the strength and weakness of the present volume. Its strength is the wide range of papers presented, offering a real overview of the state of gunpowder studies around the world with contributions from North America, Europe, India, China and Australia. Even if the papers are short, all have bibliographies and notes allowing further research and in-depth study, especially for the papers in languages other than English. My only criticism is perhaps the poor translation of some of the foreign papers, which occasionally results in some confusion about the actual point being made, a real difficulty when precise technological vocabulary is required. However, set against the range and breadth of such a volume this does seem like carping. Taken overall this is a significant volume, bringing, as it does, the study of what is arguably the most important invention of the early modern period up to date.

Robert D. Smith

Brian Cotterell and Johan Kamminga, *Mechanics of Pre-industrial Technology*, Cambridge University Press, Cambridge, 1990, paperback ed. 1992

All too often the explanations of how early artefacts were made and of how early machines worked are given in such general terms that they slide over the critical factors. Readers, however well trained in their own fields, are left with the feeling that they have not really grasped the main principles. An engineer and an archaeologist have joined forces to produce a book that aims to remove those inadequacies. Its principal objective is to demonstrate how an understanding of the principles of mechanics can help to create a better understanding of the functions of artefacts and hence of the achievements of early technology.

The range the authors cover in eleven chapters is extraordinarily wide in period, distance and complexity. After an introductory review in the first chapter of the nature of invention, theories of cultural diffusion, artefact sophistication and complexity, its second half sets out the development of mechanical science and the mechanics of material culture, with a brief survey of how mechanical problems are analysed theoretically and then tested experimentally. The next three chapters demonstrate the basic principles of mechanics, the properties of fluids and solids, and the elements of machines, a coverage that many textbooks devoted to the subject would do well to match. All three illustrate the principles by practical examples. The remaining chapters effectively reverse this procedure by using these basic principles to explain structures, stone tools and projectiles – from the spear and the boomerang to catapults – and to estimate the efficiency of different means of transport by land and water. The last chapter examines the way in which a large range of musical instruments produce their very different sounds and the way in which those sounds affect us. There follows a short epilogue, more properly a survey of the subject, with three appendices: a glossary of mechanical terms; notes on the symbols used; and some mathematical formulaes. Every chapter of the authors' exposition uses a variety of examples, from the earliest stone tools of two million years ago to the problems of the erection of statues on Easter Island, which continued until the seventeenth century, and includes representative material from most areas of the world. They rightly claim that many of the techniques and artefacts that look simple reveal quite sophisticated understanding.

In one sense, the book is anachronistic: few of those who created so many of the artefacts and who achieved so much in the development of early technology had any grasp of the ideas, concepts or the principles set forth here. But this comment is only a warning to readers not to misinterpret the authors' intentions and assume that the inventors, known and unknown, were authorities on applied mechanics. Some, of course, were; Archimedes and Heron of Alexandria are among them.

The authors' efforts to demonstrate the way in which mechanics can help demonstrate how so many of these crucially important devices were made and employed are valuable and effective. Their few failures are perhaps due to their exceeding their principal objective. For they not only review the relevant parts of mechanics and use them to explain the artefacts and machines, but also offer accounts of the development of the science of mechanics and its increasing effectiveness in explaining how the artefacts were made and how they operated. Some of this material can be distracting. An example is their discussion of the intriguing and seemingly paradoxical result relating to the frictional forces between two plates rubbed together. As they put it (p. 27):

> Although the two most fundamental statements on friction are attributed to Guillaume Amontons (1663–1705), they had already been made some 200 years earlier by Leonardo da Vinci [ref. given]. Amontons reported to the French Royal Academy that the frictional force between two sliding surfaces was proportional to the force pressing them together and independent of their contact area. The Academy accepted Amontons' first observation, but the second caused some astonishment and the senior academician de La Hire was motivated to check his experiments [ref. given]. De La Hire's experiments confirmed Amontons' findings.

They then gave the modern explanation for the apparent paradox: 'surfaces – even precision ground steel ones – only make contact over a finite number of asperities so that their actual area of contact is proportional to the force. Therefore the total area of the surface is unimportant.' But it is this paradoxical result that should receive the emphasis in a book whose readership is concerned with practical problems and their interpretation. Instead, the exposition leads the reader through the historical argument, whereas it ought to concentrate upon the actual result. The authors have let their enthusiasm for the history of mechanics lead them astray from their main objective, which is the explanation of the technology itself. The inclusion of comments on these advances in technological knowledge, and on those who made them, should be limited to those that are relevant to how the artefacts or inventions were devised or to the way these advances have been used to explain how the artefacts and machines of earlier periods operated. Discussion of earlier explanations should be confined to footnotes. There, even at some length, they would contribute insights into the history of mechanical science and engineering. Some are: rightly, the story of the apple and Newton's sudden grasp of the notion of universal gravity is in a footnote (p. 23, n. 2). However, some of the historical explanations go well beyond the needs of the book and should be eliminated: the question of the priority between Huygens and Newton over centripetal acceleration is an example (p. 33, n. 5). Moreover, the authors have relied on some very old references, notably Singer *et al.*'s *History of Technology*. That usage shows how badly we need an updated 'Singer' – a history of engineering science and technology. That is not to decry the need also for histories of the influence of society on technology and the influence of technology on society. There is a tendency to forget that these are three separate histories, no matter how closely related.

The editing and presentation are not consistent: some equations are given in general terms and then numerical values are inserted, but with others, the numbers are present within the equation. Sometimes readers will need to insert their own values, and they will then have to re-create the original equation; but that defeats the purpose of the book.

What the reviewer cannot determine is how far the authors meet their declared objective. The explanations are often demanding, not because of the authors' lack of skill in explaining them, but because what they are trying to explain is often not simple. To a one-time physicist-cum-mathematician, the explanations that he has checked are satisfying and accurate. Thus the book will undoubtedly be valuable to those whose memory needs jogging or who somehow never studied a particular part of the subject. Whether archaeologists have sufficient mathematical, mechanical and physical understanding to follow some of the derivations of the equations, understand them or use them lies outside the present reviewer's competence to decide. However, he was dismayed by the startling contrast between the sophistication of some of the concepts listed in Appendix I (glossary of mechanical terms) and Appendix II (symbols) and the elementary trigonometric equations listed in Appendix III. Those who need reminding of the definition of sine and cosine may never have heard of many of the terms in the first two appendices and will find the discussion of a subject such as fracture mechanics (pp. 135ff.) very hard indeed to follow. The book thus appears to be most useful for those instructing archaeologists, rather than as reading matter for most students of archaeology.

Other than this reservation, the criticisms appear minor compared with the authors' achievements. Not the least of these is their gathering together, from a variety of sources ancient and modern, experimental values not otherwise available. Their resort to beer as a reward to their students to obtain numerical values for some of the more esoteric constants is also to be commended. However, a word of warning to the authors, not the readers. Given the number of gaps in the information that they collected, particularly for those many processes now obsolete, they need to exercise some restraint. Future students must not be able to complain that their lecturers only have them there for the beer.

Dennis Simms

Contents of Former Volumes

* indicates out of print

FIRST ANNUAL VOLUME, 1976*

D.S.L. CARDWELL and RICHARD L. HILLS, Thermodynamics and Practical Engineering in the Nineteenth Century.
JACQUES HEYMAN, Couplet's Engineering Memoirs, 1726–33.
NORMAN A.F. SMITH, Attitudes to Roman Engineering and the Question of the Inverted Siphon.
R.A. BUCHANAN, The Promethean Revolution: Science, Technology and History.
M. DAUMAS, The History of Technology: Its Aims, its Limits, its Methods.
KEITH DAWSON, Electromagnetic Telegraphy: Early Ideas, Proposals and Apparatus.
MARIE BOAS HALL, The Strange Case of Aluminium.
G. HOLLISTER-SHORT, Leads and Lags in Late Seventeenth-Century English Technology.

SECOND ANNUAL VOLUME, 1977*

EMORY L. KEMP, Samuel Brown: Britain's Pioneer Suspension Bridge Builder.
DONALD R. HILL, The Banū Mūsá and their 'Book of Ingenious Devices'.
J.F. CAVE, A Note on Roman Metal Turning.
J.A. GARCIA-DIEGO, Old Dams in Extremadura.
G. HOLLISTER-SHORT, The Vocabulary of Technology.
RICHARD L. HILLS, Museums, History and Working Machines.
DENIS SMITH, The Use of Models in Nineteenth-Century British Suspension Bridge Design.
NORMAN A.F. SMITH, The Origins of the Water Turbine and the Invention of its Name.

THIRD ANNUAL VOLUME, 1978*

JACK SIMMONS, Technology in History.
R.A. BUCHANAN, History of Technology in the Teaching of History.
P.B. MORICE, The Role of History in a Civil Engineering Course.
JOYCE BROWN, Sir Proby Cautley (1802–71), a Pioneer of Indian Irrigation.
A. RUPERT HALL, On Knowing, and Knowing How to ...
FRANK D. PRAGER, Vitruvius and the Elevated Aqueducts.
JAMES A. RUFFNER, Two Problems in Fuel Technology.
JOHN C. SCOTT, The Historical Development of Theories of Wave-Calming Oil.

FOURTH ANNUAL VOLUME, 1979*

P.S. BARDELL, Some Aspects of the History of Journal Bearings and their Lubrication.
K.R. FAIRCLOUGH, The Waltham Pound Lock.
ROBERT FRIEDEL, Parkesine and Celluloid: The Failure and Success of the First Modern Plastic.

FIFTH ANNUAL VOLUME, 1980*

SIXTH ANNUAL VOLUME, 1981*

SEVENTH ANNUAL VOLUME, 1982*

EIGHTH ANNUAL VOLUME, 1983*